The Insiders' Guide to
ISO 9001:2008

Other Books from Paton Professional:

The Corrective Action Handbook
By Denise Robitaille

Document Control
By Denise Robitaille

ISO 9001 in Plain English
By Craig Cochran

The Management Review Handbook
By Denise Robitaille

Managing Customer-Supplier Relationships
By Denise Robitaille

The Preventive Action Handbook
By Denise Robitaille

Root Cause Analysis
By Denise Robitaille

For more information on these and other Paton Professional books, visit www.patonprofessional.com

The Insiders' Guide to ISO 9001:2008

Lorri Hunt

Denise Robitaille

Craig Williams

PROFESSIONAL

Chico, California

Most Paton Professional books are available at quantity discounts when purchased in bulk. For more information, contact:

Paton Professional
A division of Paton Press LLC
P.O. Box 44
Chico, CA 95927-0044
Telephone: (530) 342-5480
Fax: (530) 342-5471
E-mail: *books@patonprofessional.com*
Web: *www.patonprofessional.com*

12 11 10 09 08   5 4 3 2 1

ISBN 978-1-932828-23-8

Library of Congress Cataloging-in-Publication Data
Hunt, Lorri, 1966-
The insiders' guide to ISO 9001:2008 / Lorri Hunt, Denise Robitaille, and Craig Williams.
   p. cm.
Includes index.
ISBN 978-1-932828-23-8
1. ISO 9001 Standard. 2. Quality assurance. 3. Quality control. I. Robitaille, Denise E. II. Williams, Craig, 1967- III. Title.
TS156.6.H86 2008
658.4'013--dc22
                    2008047350

Staff
Publisher: Scott M. Paton
Editor: Laura Smith
Book design: Caylen Balmain

Dedications

Lorri Hunt

To my mom, Connie, and sisters Gwen and Heather whose love and support over the last few years have seen me through a victory over breast cancer and career changes that have led me to doing the work I love best. I could not have done it without you.

Denise Robitaille

To the many teachers at St. Anthony's and at the University of Massachusetts at Amherst who taught me the value of respect, conviction, diversity, and richness of language—lessons without which this book and my work would not be possible.

Craig Williams

To my wife, Caroline, and my boys Alex and Matt who have endured my travels and still make home the best place in the world to be. To my parents, who have always had the faith in me to help me dream. To my brother and sister who continue to keep me grounded in who I am. All of their love is invaluable.

To my colleagues at Eaton Corp. who have continued to display their commitment to quality through sponsoring my work with ISO/TC 176. It is truly appreciated.

CONTENTS

Acknowledgments

We'd like to recognize our colleagues and friends on the U.S. TAG and ISO/TC 176. These dedicated individuals share our commitment to the development and support of international standards that benefit the global community. We also want to convey our sincere gratitude to them for enriching our work and our lives.

PART 1

INTRODUCTION

This book provides an in-depth look at the 2008 revision of ISO 9001. We will explain the changes found in the standard and discuss the process, the rationale, and the revision's effect on users. We will also provide insights as to why the changes were made from the perspective of technical experts from the United States who were involved in editing ISO 9001:2008. This book will provide first-hand knowledge on why each change was made and provide guidance on possible modifications you may choose to make to your own quality management system (QMS).

The technical experts of ISO/TC 176, who write and update the ISO 9000 series of standards, have been vigilant in gathering data from user communities and reviewing market feedback while updating ISO 9001. The results of this data analysis yielded two key decisions: First, ISO 9001:2008 would be an amendment with only minor changes, and, second, ISO 9001:2015 (an approximate publication date for ISO 9001's next revision) would be a more extensive revision that will address global and technological changes in the marketplace. Thus, all the feedback deemed appropriate for the 2008 amendment was addressed in this year's version. The more complex issues were tabled for the next revision.

It's important to affirm that the amendment doesn't add any new requirements, but it would be unfortunate to dismiss it as unimportant. The changes that have been made will increase the utility of the standard and will better serve the marketplace. Through clear articulation of requirements and elimination of ambiguities, organizations have an opportunity to better grasp the intent of ISO 9001's existing requirements to experience greater value from their implementations.

The clarification of existing requirements also improves the auditing process by ensuring that there is consensus as to their meaning. As the market for the ISO 9000 family of standards has grown and the number of sector-specific spin-offs it has spawned increase, the need for consistent application of the requirements is proportionately amplified. Geographic constraints and language barriers engendered by the global economy mandate unambiguous, universally applicable standards. The integrity of conformity assessment relies on standards to be adequately clear to ensure uniform criteria for certification.

Thus, the amendment to ISO 9001 is specifically intended to remove the obstacles to comprehension and implementation of the standard.

The first part of this book highlights some of the changes that could require an enhancement to your QMS, what the transition plan for the amendment entails, provides information about the International Organization for Standardization (ISO) drafting process, the inputs that fueled the amendment, the effect on sector-specific standards, and some insight into what the future holds. The second part of this book addresses each clause or subclause of the standard. Even if the changes don't appear to affect your organization, this revision provides an excellent opportunity for you to review and improve your QMS. There are several beneficial outputs to be realized from a review of your QMS. You will be able to identify:

■ Problems

■ Areas where you have misunderstood and misapplied requirements

■ Opportunities to improve processes based on a clearer understanding of a requirement

At the end of the book you'll find an appendix that we lovingly refer to as the "ISO Geek Speak Annex." We noticed throughout the writing process that our language and, consequently, our book is peppered with acronyms, abbreviations, and numbered designations for committees, documents, task groups, and projects. Although they're almost as familiar to us as our own names, we've had to concede that it's just a lot of gibberish to the casual listener. Acknowledging that this will not relieve us of the title "ISO Geeks," we offer this appendix in an attempt to demonstrate that our ISO alphabet soup really does make sense. We also thought it would increase your understanding of the revision process and ultimately the intent of ISO 9001.

We hope this book will serve you well and wish you luck with the review process.

—Craig, Denise, and Lorri

CHAPTER 1

UNDERSTANDING THE CHANGES

A s users get their first glimpse of ISO 9001:2008, the question on everyone's mind is, "What, if anything, will our organization need to do differently?" ISO 9001:2008 focuses on changes that organizations might make to better comply with the spirit of the standard without adding, deleting, or altering its requirements. It should not result in an extensive change to existing quality management systems (QMS). The changes are minor in nature and address such issues as the need for clarification, greater consistency, resolution of perceived ambiguities, and improved compatibility with ISO 14001, which relates to environmental management systems. The rationale for these changes is discussed in greater detail in chapter 3.

What does this mean for users?

Requirements in the standard are frequently referred to as "shalls." For the purpose of this amendment, ISO 9001:2008 provides improvements for users without adding to or removing any of the "shalls."

ISO/TC 176, the technical committee responsible for updating the ISO 9000 standard series, was careful not to make change just for the sake of making change. This was especially true when it came to editorial changes, which could have created the false impression that there was a change in requirements, carrying greater significance than was intended. In some instances, when the committee members couldn't come to a consensus in determining if a change added or deleted a requirement, they opted to retain the existing text. It was decided to err on the side of caution rather than to contribute to any misunderstanding in the marketplace.

With that said, some of the changes that have been incorporated into this amendment have been or have the potential for being perceived differently than what the technical experts intended when they were drafting the document. To minimize this, the committee made some very deliberate word choices. Some of the changes appear to be mere nuances of similar concepts. However, they often carried the greatest potential for incorrect user interpretation. Other changes increased understanding across a broader range of product types, including service organizations. Finally some changes to specific clauses were made based on the 2004 International User Feedback Survey. This survey was conducted after the publication of ISO 9001:2000 and invited respondents to identify areas they most wanted to see improved.

Although it's already been mentioned that all changes are described in full in the second part of this book, this chapter will highlight those changes that seem to have the most importance and which will probably be of greatest concern to the average user. What follows is a discussion of the most significant subject matters addressed in ISO 9001:2008.

OUTSOURCING

The term "outsourcing" has been discussed at length since the publication of ISO 9001:2000. In fact, it spurred guidance documents as well as articles to explain its use in clause 4.1, as well as its relation to the purchasing process in clause 7.4.1. In an attempt to clarify the intent, notes were added to clause 4.1 that require the organization to identify processes to be completed by an external party and the factors influencing the nature and extent of control over these processes. The added text is in the form of notes, thereby serving to provide clarification—not additional requirements.

What to consider

In reviewing the new note that clarifies the relationship of outsourcing to the organization's purchasing process, it's important to remember that this is just one of the methods that can be used to control the process. If the organization outsources a process to another party, such as a corporate headquarters, the purchasing process would not necessarily be sufficient to control the QMS implications of that process, so other methods would need to be established. Therefore, organizations need to analyze each process separately to determine the necessary controls.

DOCUMENTATION

Clause 4.2, Documentation requirements, was identified in the 2004 International User Feedback Survey as an area that needed improvement. This was also an area of the standard where the technical experts were extremely careful to avoid adding or deleting requirements. This was due to the potential of requiring organizations to make extreme modifications to their QMS and to their documentation in particular.

The technical experts focused on clarifying what documentation was needed for the QMS by modifying clause 4.2.1, General. Clarification was made by structuring this clause such that it did not identify records as a separate type of document. Additionally it was changed to indicate that a documented procedure required by ISO 9001 that is established, documented, implemented, and maintained can be a stand-alone document that addresses more than one procedure or be covered by more than one document.

The changes made to clause 4.2.3, Control of documents, and 4.2.4, Control of records, were focused on improving the compatibility between ISO 9001 and ISO 14001.

What to consider

Organizations should evaluate whether they need to make changes to the structure of any of their procedures based on the clarification that procedures can be combined or stand-alone. These changes should focus on the needs of the organization and help to improve its documentation system.

MANAGEMENT REPRESENTATIVE

The changes in clause 5.5.2 aren't significant if you count the number of words that changed. The only addition was to require that the management representative be a member of the "organization's" management. Not many words were added, but there may be many companies that see this as a significant change. Primarily, there is a debate as to whether a company (particularly smaller companies) can use a contracted person to be the management representative based on this wording change. Although some individuals may interpret this to mean that the management representative must be a full-time employee of the organization, the requirement is that the management representative needs to be part of the organization's management. This means a contracted person could serve in this role as long as he or she is also considered as part of management and has been assigned the necessary authority and responsibility.

What to consider

When analyzing this change, organizations need to consider whether they have a person assigned as management representative who meets the criteria as explained above. Based on the results of this analysis, you may need to more clearly document the relationship of the management representative to the organization.

COMPETENCE OF EMPLOYEES

A note has been added to clause 6.2.1 stating that "conformity to product requirements can be affected directly or indirectly by personnel performing any task within the quality management system." This note was made to clarify that the conformity of product requirements can be affected by someone who builds the product (directly) or a buyer who is responsible for purchasing materials that are to be used in building the product (indirectly).

Why is there the potential for this change to be perceived as new or different? In the past, some organizations took a very specific viewpoint that only some of their employees affected the conformity of product to requirements. These were typically the employees who built the product or delivered the service. In doing this, they at times eliminated some employees who indirectly affect that conformity. This means that some organizations did not apply this clause as thoroughly as they should have. This note doesn't change the original intention of the requirements of clause 6.2.1, which is that personnel performing work affecting conformity to product requirements shall be competent. The emphasis is simply stressing that the effect can be direct or indirect.

What to consider

Organizations need to consider if they have applied this clause to employees who affect conformity to product requirements indirectly as well as directly. If gaps are identified, competence for these employees should be established just as they are for employees who directly affect conformity.

DESIGN REVIEW, VERIFICATION, AND VALIDATION

One of the clauses that users have struggled with in the past is 7.3, Design and development. ISO/TC 176 received many comments regarding the clarification of the design activities of review, verification, and validation. Specifically, users were confused on how these activities relate to each other and whether they could be conducted by themselves or simultaneously. Although the technical experts determined that ISO 9000 adequately defines "design" and "development," it was apparent that users still did not understand their application. For that reason, a note was added to clause 7.3.1 that indicates that these activities have distinct purposes but can be conducted by themselves or in combination with one another.

What to consider

Organizations that include design and development in their QMS should evaluate if they have been conducting the activities of review, verification, and validation based on an understanding of what they thought the standard required rather than based on an infrastructure that works for the organization. Based on an improved understanding of the requirements, the organization may find that it can make improvements to streamline these activities.

EQUIPMENT

Throughout the standard the word "devices" was changed to "equipment." How did the technical experts come to this conclusion? During the creation of the design specification for the amendment, one of the areas requested for clarification was the term "device." As background, the word "device" was used in ISO 9001:2000 to address devices that were not considered equipment, such as checklists.

In reviewing potential solutions and terminology as it existed in ISO 9000:2005, it was determined that the term "measuring equipment" included "measuring instruments," which also could include devices. It's also believed that the current first sentence, which indicates that the organization "shall determine the monitoring and measurement to be undertaken" addresses the use of devices.

The potential for misinterpretation exists in that some organizations may see this as a reduced requirement or that devices are no longer covered by this clause. Again, this does not change requirements, but clarifies for users what monitoring and measuring equipment includes.

What to consider

Does the use of the word "equipment" instead of "device" require your organization to make any clarifying changes to your QMS? If you determine that you have not been including all of the required types of measuring equipment, the organization needs to identify these items and apply each of the requirements in clause 7.6.

MONITORING AND MEASUREMENT OF PROCESSES

Since ISO 9001:2000 was published users have been confused about how the clause on monitoring and measuring processes is related to monitoring and measuring product. This confusion can be attributed to the phrase at the end of the clause 8.2.3 which indicates that actions taken shall be to "ensure conformity of the product." This clause has been removed in ISO 9001:2008 to eliminate this confusion. In addition, a note has been added that indicates that the monitoring and measuring of processes should be appropriate to the effect on the conformity to product requirements and on the effectiveness of the QMS.

What to consider

When reviewing this change, the organization needs to consider adjusting the processes it monitors and measures and the methods it uses.

CONTROL OF NONCONFORMING PRODUCT

Clause 8.3 was revised to make it more relevant and user-friendly for service organizations. From the time that ISO 9001 was originally published, there has always been difficulty in applying clause 8.3 to service organizations. This is because it's almost impossible to identify a potential nonconformity prior to delivery to the customer, which results in the organization almost always being in a corrective action mode.

The technical experts evaluated several options to make this clause more universally applicable. The last paragraph from ISO 9001:2000 was added as one of the ways that an organization can deal with nonconforming product: "When nonconforming product is detected after delivery or use has started, the organization shall take action appropriate to the effects, or potential effects, of the nonconformity." By implementing this approach, an organization can take action appropriate to the effects, or potential effects, of the nonconformity when nonconforming product is detected after delivery.

Another small change was to add the words "where applicable" as the lead-in to the text that addresses nonconforming product. This is to emphasize that where an organization can apply the requirements, it should. This clarification was made to stress that some organizations—based on the nature of their product type—may not be able to apply all of the methods for dealing with nonconforming product.

ISO/TC 176 thoroughly examined the potential for an organization not to apply certain aspects of the clause due to the revisions. It determined that the words "where applicable" puts the onus on organizations to apply the requirements where they can and that the potential for misinterpreting this clause is far less than the benefits it derives for organizations that have struggled to comply.

What to consider

Organizations, especially those in the service industry, should consider whether they can make enhancements to how they are controlling nonconforming product based on the inclusion of taking action appropriate to the effects of the nonconformity as a method of controlling nonconforming product. Any enhancements made due to this clarification should be included in the organization's documented procedures.

CORRECTIVE AND PREVENTIVE ACTION

Clauses 8.5.2 and 8.5.3 of ISO 9001:2000 required the review of corrective and preventive actions. The technical experts evaluated the consistency between these clauses and their counterparts in ISO 14001:2004 and ISO 19011. To improve the consistency, the technical experts decided to add the words "effectiveness of the" to the review of corrective and preventive actions in ISO 9001:2008.

In making this change the technical experts considered the term "review" in ISO 9000:2005. It was determined that the term "review" includes consideration of the effectiveness of an activity taken, so this change was not considered an additional requirement.

Users could perceive this as a change in requirements. However, effectiveness was always considered to be part of the activities of review for ISO 9001.

What to consider

Organizations need to analyze their QMS to ensure that when they are reviewing corrective and preventive actions they are considering the effectiveness of the actions taken. If your organization has not included a review of effectiveness, this should be incorporated into your QMS and the respective procedures updated.

STATUTORY AND REGULATORY

One of the changes made throughout the standard was the change from "statutory" to "statutory and regulatory." This includes changes made to clause 0.1, General, in the Introduction section, as well as 1.1, General, and 1.2, Application, in the Scope section of ISO 9001:2008. There was considerable discussion during drafting the amendment about the distinction between the terms "statutory" and "regulatory." These terms may give some countries the impression of change due to interpretation. In the United States it's perfectly acceptable and appropriate to talk in terms of statutory and regulatory requirements as potentially different requirements. However, in most of the world these two terms are synonymous and are further complicated when ISO 14001 talks in terms of "legal" requirements. Although these items may seem trivial, the conversation they generate is very passionate and focused on a uniform and global understanding of the documents, particularly for tens of thousands of users who are not native English speakers.

To further clarify the meaning of "statutory and regulatory," a note has been added that explains what is meant by these type of requirements as well as legal requirements.

What to consider

Because the phrase "statutory and regulatory" already existed in ISO 9001:2000 clause 7.2.1, ISO 9001:2008-compliant organizations shouldn't have to change their QMS. However, organizations that have taken a minimal compliance approach in this area may need to consider whether the inclusion of "statutory" in the additional clauses changes any of the requirements they need to consider when determining the scope and application of their QMS.

OUTPUT MATTERS

There was an additional significant consideration during the drafting of ISO 9001:2008. There have been organizations registered to ISO 9001 that did not meet customer expectations or that produced products that didn't meet requirements. This has created an undercurrent that threatens to undermine the credibility of the standard. In recent years, this has given rise among members of the user community, technical experts, consumers, and quality professionals of the concept that "output matters." It's not good enough for an organization to have an ISO 9001-registered QMS. It's of equal or greater importance that the *output of that organization meets customer expectations.*

While ISO 9001:2008 was being developed, the ISO/International Accreditation Forum (IAF) Conformity Assessment Liaison Group (CALG) was reviewing ways to promote the concept that "output matters." The IAF provided comments on ISO 9001 to the technical experts participating in the review of ISO 9001 relating to changes that could be made to emphasize this point. In most cases, the technical experts determined the wording changes to go beyond the scope of the amendment. Comments that could not be incorporated into the amendment have been retained and will be considered in the next full revision of ISO 9001. However, a change was made to clause 0.2, Process approach, to clarify that the processes established in the system must produce the desired outcome. Although the wording change doesn't add a requirement, it does emphasize that the output of the QMS is a component that should not be ignored.

What to consider

What does this mean to the user community? Because this isn't a change in requirement, it's not necessary for an organization to do anything differently. However, the change in the text should not be dismissed or trivialized. It should be the first step in a concerted effort to ensure that the integrity of ISO 9001 is retained. Therefore, if your organization has struggled with producing products or delivering services that meet customer requirements, it should leverage ISO 9001:2008 to make improvements to its QMS.

MAKING CHANGE WITHOUT CHANGING REQUIREMENTS

It's important to remember that the goal of this amendment is to make improvements to the existing standard without adding or deleting requirements. How were the technical experts who produced the amendment able to make clarifications without changing requirements?

One significant mechanism was the use of notes throughout the standard. Section 0.1, General, indicates that information marked "Note" is for guidance in understanding or clarifying the associated requirement. By adding or expanding notes in the standard, requirements could be clarified without changing the requirement itself.

The technical experts also considered ISO 9000, Quality management systems—Fundamentals and vocabulary, to determine if changes were needed to ISO 9001. ISO 9000 is the terminology document for the ISO 9000 family of standards. It defines terms that assist an organization in understanding the application of ISO 9001.

WHAT'S NEXT?

The remaining changes in ISO 9001:2008 provide additional clarification or improve consistency within the document. Part two of this book provides an analysis of each of these changes. It also provides a review of existing requirements so that organizations can conduct an in-depth analysis of its QMS.

As with many of the changes that were made in this amendment, it's important to examine the changes holistically. There are several steps users should take to avoid misinterpreting the standard:

- Look at the change in the context of the clause in its entirety. If you look only at a specific sentence, there is the potential for misunderstanding the intent.
- Refer to ISO 9000:2005. ISO 9000 is the terminology document for ISO 9001. It defines terms that are critical to the understanding of requirements.
- Finally, remember the intent of the amendment was not to add or delete requirements, thereby avoiding the need for any significant change for organizations. However, some users might find that based on their improved understanding of the requirements, they might need to make slight adjustments. If organizations keep this in mind when considering change, they will improve the quality of changes they are making to their QMS.

The various rationales we've just discussed represent the contextual rationales for the changes. We'll spend more time talking about the standards-writing process in chapter 5. For now, it's adequate to note that it takes several years from inception to final release for a standard to be amended.

CHAPTER 2

WHAT NOT TO WORRY ABOUT

t's important to not over-interpret the changes in ISO 9001:2008. Remember, they are not new requirements. Be sure that you comprehend the significance of the changes before expending time and resources changing your documentation or processes.

The changes from ISO 9001:2000 to ISO 9001:2008 won't result in the worldwide transition projects that arose from the 2000 revision of the standard. This means no added procedure requirements and no transition audit.

To emphasize this point, at its May 2008 meeting ISO/TC 176 adopted a resolution stating: "Noting that ISO 9001:2008 does not provide new requirements, ISO/TC 176 resolves that:

■ ISO 9001:2008 has been developed in order to introduce clarifications to the existing requirements of ISO 9001:2000 and changes that are intended to improve compatibility with ISO 14001:2004. ISO 9001:2008 does not introduce additional requirements nor does it change the intent of the ISO 9001:2000 standard.

■ Certification to ISO 9001:2008 is not an upgrade, and organizations that are certified to ISO 9001:2000 should be afforded the same status as those who have already received a new certificate to ISO 9001:2008."

The International Accreditation Forum (IAF) recently published the transition plan for ISO 9001:2008, which confirmed the resolution taken by ISO TC/176. The following key points are addressed in the paper:

■ *No certifications to ISO 9001:2008 until it is published.* This means organizations cannot certify to any draft versions or balloted versions approved by ISO/TC 176 until it is officially published by ISO.

■ *No certifications to ISO 9001:2000 after twelve months.* Twelve months after ISO 9001:2008 is published, no new certificates will be issued to ISO 9001:2000. This means that if an organization chooses to pursue its current path of ISO 9001:2000 certification, it can continue as long as it is within twelve months of the publication of ISO 9001:2008.

■ *Organizations have twenty-four months to meet the requirements of ISO 9001:2008.* This assessment will be performed at the time of a regular surveillance or recertification audit. Twenty-four months after publication of the updated standard, ISO 9001:2000 will be withdrawn and existing certificates to it will no longer be considered valid.

Keeping this in mind, organizations will need to consult their registrar as they incorporate their specific requirements into their transition to ISO 9001:2008.

Why is the transition different than for ISO 9001:2000? You'll recall that the changes from ISO 9001:1994 to ISO 9001:2000 were considerable. The work to ensure conformance and effective implementation during the transition was significant, with proportionate costs and labor expenditures. It was a greatly improved standard, but complying to it was a major endeavor.

Ensuring conformance to ISO 9001:2008 shouldn't create similar upheaval in your organization. If you have a robust system, you shouldn't have anything to worry about.

You may, however, decide that certain changes may be appropriate to more fully comply with the spirit of the updated standard. These changes could encompass documentation, implementation, or both.

There are two scenarios that would indicate the need for change. The first case involves a requirement that has been misunderstood and improperly implemented. The authors of ISO 9001:2000 recognized that some of its language was ambiguous and could lead to the misapplication of a requirement. Bearing in mind that part of the intent of the changes found in ISO 9001:2008 is to address those ambiguities, there is significant likelihood that organizations may make changes in their policies and practices based on a better understanding of the requirements.

The second case involves correcting the implementation of a practice that did not provide appreciable benefit due to lack of comprehension of the intent of the previous version. Here organizations have the opportunity to make changes that will improve an existing process to experience greater value.

In both cases, this is good news. It means that the changes you make will strengthen the integrity of your QMS.

ISO 9001:2008 presents a great opportunity for organizations to assess their systems. Part two of this book contains worksheets that describe the changes and their significance. Each worksheet has a description of the purpose of the requirement and a checklist of features. Part two also includes worksheets for all the clauses, even those that have no changes. A complete assessment will yield not only information on what needs to be addressed as a consequence of the amendment but will also identify any other areas requiring corrective action or improvement.

CHAPTER 3

AMENDMENT VS. REVISION

I t's important to discuss the scoping process that was used in determining what was to be changed in the ISO 9001:2008 drafting process. For ease of reading, wherever possible we will refrain from using the ISO-speak that often floats around in the standards world while explaining the drafting process. It's a world of numbered documents and tedious text bordering on legalese that can put even the most devout quality professional to sleep.

Chapter 5 will discuss the standard's revision process in detail. Conversely, this chapter will cover the aspects of the process that are pertinent to understanding what, how, and why things changed—and, conversely, why some things didn't change.

Once the determination is made that a revision to a standard is warranted, a document is created to specify its scope. The document is called a design specification. In the case of ISO 9001:2008, the document was titled, "N730: Design Specification for an Amendment to ISO 9001:2000." The design specification set boundaries for the technical experts to use so that their work would reflect the planned changes in regard to both nature and extent.

Although documentation references both the terms "amendment" and "revision," there is no formal distinction in the International Organization for Standardization (ISO) world between the two terms. In this case, the nature of the amendment was included in the design specification. The key points of the design specification were:

- The overall clause structure of the document was to remain unchanged.
- No new requirements would be added and no current requirements deleted.
- Movement of requirements between clauses would only be allowed where it did not change, add, or delete a requirement.

Consequently, the changes you will see in the document will not cause wholesale changes in well-established systems.

The review and revision processes employed prior to the drafting of ISO 9001:2008 yielded significant benefits.

The review process allowed the user community to affirm the continued relevance and usability of the standard. ISO 9001 remains the world's premier quality management system (QMS) model; it's used by about one million organizations around the globe. The user community let the technical experts know through multiple feedback conduits that it is engaged in the dialogue about its application now and in the future.

Another benefit arose out of a constraint in the design specification that required dismissing comments that were determined to be outside of the defined scope. The technical experts needed to provide clarifications identified in the design specification without breaking any of the previously mentioned rules. Throughout the drafting process, there were ample opportunities for experts from around the world to provide feedback into the process and request areas that they wanted to be addressed for revision. In many cases, the design specification process weeded out suggestions that would have gone beyond the agreed-upon scope.

Many good ideas that were raised exceeded the scope of the amendment. What happened to these ideas? The ideas and recommendations were maintained as part of the formal records during the drafting process. The technical experts were diligent in reviewing all of the suggestions and ensuring that feedback from stakeholders was not lost. When the next drafting committee is formed, these records will constitute a portion of the input to the process. It's anticipated that some may very well find their way into a future revision.

WHY THE UPDATE?

There has been much discussion in the marketplace about the need and the timing of this amendment. What sparks the need for a revision and who decides when? The answers are rooted in ISO's Guide 72, which requires committees that develop management standards to conduct a review of published standards every five years. The review helps to ensure ongoing relevance to the user community and justifies changes.

ISO/TC 176 identified multiple rationales for the changes it made. These rationales helped it to remain within the scope of the design specification without adding requirements. Therefore, the changes had to fit into one of the following categories:

- Clarification
- Consistency
- Compatibility with ISO 14001
- Resolve ambiguities
- Increase user comprehension and applicability
- Continued relevance
- Eliminate the need for interpretation documents

CLARIFICATION

Clarification was needed in several areas to address requests for interpretation that were received by various international groups. These requests resulted in multiple official interpretation documents that will be described in greater detail later in this chapter. Clarifications often improve the usability of the standard without adding new requirements.

CONSISTENCY

Another focus of the amendment process was to improve consistency within ISO 9001 itself. One of the challenges in writing standards is to resist the occasional inclination to use creative language. Terminology must be clear, correct, and consistent. Therefore, it's critical that the same phrase be used in exactly the same manner throughout the standard. An example was using the same phrase to articulate the requirement that a documented procedure be established, implemented, and maintained, rather than using different wording in each of the clauses requiring a documented procedure.

There's a practical benefit to be realized from improved consistency. It's rooted in the purpose and nature of the standard. ISO 9001 is utilized globally as the benchmark criteria to qualify suppliers. Ensuring that the requirements of the standard are consistently applied is more than an esoteric exercise. It relates directly to the confidence organizations may have that suppliers around the globe are playing by the same rules. They can be assured that processes relating to requirements such as controlling documentation, understanding specifications, design, verification of product, and addressing problems are conducted consistently regardless of geography or industry.

COMPATIBILITY WITH ISO 14001

To date, the revision cycles for ISO 14001 and ISO 9001 were not synchronized. This has resulted in the respective technical committees making changes to their documents that, in turn, must be addressed by the other technical committee in subsequent revision cycles. For example, during this amendment the technical experts looked at the requirements for controlling documents and controlling records as areas for improved compatibility between ISO 9001 and ISO 14001.

RESOLVE AMBIGUITIES

As standard bearers and technical experts of lofty documents such as ISO 9001 and ISO 9004, it pains us to concede that even quality-minded folks might create errors and ambiguities that can drive the user community crazy. Therefore, part of the process and rationale for the amendment was to address and clarify some of the errors in the language of the documents. Ambiguity creates the risk that requirements could be interpreted in different ways by users and registrars.

Some of the errors are the standard things that you can expect in any document, including unintended omissions or errors that editors miss. Sometimes, however, the use of certain words can create ambiguities if they aren't used consistently throughout the document. For example, consider words that are both a verb and a noun, such as "document." There can be some confusion in its use as to whether it means the process of documenting or a physical or electronic document. The word "review" can be seen as a verb or as an event. The interchangeable use of words like "identify," "identity," and "identification" can really cause confusion.

Throughout the drafting process there are always key phrases that are used in ways that may go right past the casual observer. One example is of two phrases that may seem interchangeable to the user but have very specific meanings in drafting are the phrases "as appropriate" and "as applicable." "As appropriate" is used to mean that the requirement is applied at the discretion of the organization to meet the minimum intent. "As applicable" is used to mean, "If it *can* be applied, the requirement *shall* be applied." This is an important distinction. It can have serious implications to the user.

All of this is further complicated by the fact that the document is translated into the myriad of languages for almost one million user communities around the world. The translators are stuck with the onerous task of trying to decide— sometimes only from context—which form of the word is meant.

INCREASE USER COMPREHENSION AND APPLICABILITY

ISO/TC 176 made significant improvements to the language describing the application of ISO 9001 to all product types. However, some areas of the standard continued to be difficult to apply to all organizations. Specifically, service organizations continued to struggle with the application of various requirements in clauses 7, Product realization, and 8.3, Control of nonconforming product. The amended standard takes additional steps to improve this application without reducing or adding requirements.

CONTINUED RELEVANCE

Not only is it important to determine if the standard is still relevant, it's necessary to ensure that it remains current. In today's world, technology and methods are changing at a rapid pace.

When ISO/TC 176 conducted its systematic review of ISO 9001, participating countries had the option to confirm the standard as-is, amend, revise, or withdraw it. Many countries based their decisions and subsequent votes on feedback from the user community. At the time of the justification study, some organizations had only recently transitioned to ISO 9001:2000. Additionally, organizations that had implemented a sector-specific version of ISO 9001, such as ISO/TS 16949 or AS9100, had been required to transition to updated versions of the standards in even shorter periods of time. Many organizations were demanding a "cooling off" period prior to a full-blown revision to ISO 9001. Not only were organizations not looking for change, they were also looking for reassurance that the change would happen no sooner than 2008. This amendment allows the standard's users to fully develop their systems without worrying about a repeat of the upheaval that some experienced when ISO 9001:2000 was published. It also allows the technical experts the time they need to thoroughly address and properly react to the changes that will affect the continued utility of the standard.

ELIMINATE THE NEED FOR INTERPRETATION DOCUMENTS

In November of 2003 ISO/TC 176 established a process for managing requests for official interpretations from the user community. This has resulted in the development and publication of more than fifty sanctioned ISO interpretations relating to various requirements found in ISO 9001.

One of the driving forces behind the 2008 amendment to ISO 9001 was the need to address the official sanctioned interpretations to the 2000 version of the standard and potentially eliminate the need for them to simultaneously exist.

ISO/TC 176 developed a process in which official member bodies and liaison organizations could request an interpretation on a requirement in the standard that was unclear to them. This process was developed at the demand of the user community, which indicated a need for an official group to review interpretations internationally.

For submitted requests, the working group on interpretations (WGI) determined if there was an interpretation issue, and, if so, an official response was prepared, balloted, and provided back to the requester. In some cases, the working group couldn't come to agreement on an official position either due to a lack of consensus on an official interpretation or disagreement on whether an interpretation was even needed.

For the purpose of the amendment, the WGI submitted a list of interpretations to be considered during the drafting process. It included sanctioned interpretations and unresolved interpretation items. When a change could be made to ISO

9001 without adding or deleting requirements, the interpretation issues were addressed.

By incorporating text to clarify interpretation issues, the technical experts were moving toward meeting one of the original goals identified in the design specification: To eliminate the need for interpretations, as well as provide clarification to users.

At the time of this writing, it's still unclear as to whether the official sanctioned interpretations will be eliminated, but it was ISO/TC 176's goal that this would happen. A group of the technical experts from the drafting task force will prepare a report after the publication of ISO 9001:2008 making recommendations on the disposition of these interpretations.

What is the value of eliminating the interpretations? Anecdotal evidence suggests that awareness of the existence of the interpretations documents is uneven at best. Some users may be aware that the interpretations exist, but don't completely understand how to utilize them to ensure conformance and effectiveness of implemented requirements. Finally, there is no requirement that the interpretations be used by organizations or by registrars conducting third-party audits. This only serves to further exacerbate the fallout from inconsistent interpretation and the resulting disparate application. All of this increases the potential for organizations to have significant differences in how they implement their requirements.

It has always been the goal of ISO/TC 176 that ISO 9001 be used as a stand-alone document. By eliminating the need for an official interpretation, the technical experts have successfully addressed an issue and responded to customer feedback regarding their product.

CHAPTER 4

SECTOR-SPECIFIC STANDARDS

SO 9001 is very generic. The document is intended to be equally valid in widely disparate types of companies; this is one of its greatest strengths but also one of its greatest perceived weaknesses.

To address the unique requirements of various industries, sector-specific standards that are derived from or supplement ISO 9001 have been published. Of these, the most widely known are ISO/TS 16949 for automotive suppliers, ISO 13485 for medical device manufacturers, and AS9100 for aerospace manufacturers. They are based on the solid foundation of ISO 9001 and use it as a framework for additional industry-specific requirements.

There is certainly a range of opinions about the proliferation of sector-specific standards. Proponents cite the problems that arise as the catalyst of multiple sector-specific standards with varying complexity, addressing diverse industries and sometimes with conflicting requirements challenge their best efforts to maintain a quality management system (QMS) that is effective and beneficial.

Sector-specific standards have benefits and drawbacks. Most prominent among the benefits is the fact that they can be incorporated into an existing ISO 9001-registered QMS. Properly implemented and well-integrated sector-specific standards can streamline systems for organizations that are required to conform to requirements from multiple industries. This facilitates activities such as supplier qualification and monitoring customer feedback.

Some sector-specific standards have additional requirements that can have applicability beyond the industry that originally spawned them. For example, automotive suppliers registered to ISO/TS 16949 deal with the production-part approval process (PPAP) and failure mode and effects analysis (FMEA). Organizations that sell within the medical devices industry (addressed by ISO 13485) have tiered requirements based on whether a device is implantable. Purveyors to the telecommunications industry (TL 9000) have varied requirements embedded in the standard to address hardware, software, and service providers. There are similar examples for the other half-dozen or so sector-specific standards. Although the requirements are appropriate and reasonable, it's important to recognize the challenges and constraints they present for users.

There are two disadvantages to sector-specific standards. First, their proliferation might dilute the value of ISO 9001 in the marketplace. If there are too many sector-specifics, there is a concern that ISO 9001 will be viewed as a lesser standard due to its generic nature. This view ignores the universality that has proven to be ISO 9001's hallmark.

Second, an over-proliferation of standards may cause companies to incur unnecessary costs as they attempt to serve multiple industries by carrying many different certifications—for medical, telecom, aerospace, automotive, or other sectors. This doesn't even include additional certifications to standards such as ISO 17025 and ISO 14001, which address test labs and environmental concerns. It can be daunting for an organization selling to several of these markets to develop a QMS that incorporates the requirements of multiple sector-specific standards effectively and efficiently. Compliance to multiple standards can become unwieldy. The technical experts of ISO/ TC 176 recognized that these standards are squarely founded on ISO 9001.

Therefore, revisions must contain language that minimizes potential conflicts with the sector-specific standards. They also needed to ensure that they didn't include requirements or language that placed unnecessary burdens on some industries for which a particular requirement might be inappropriate or irrelevant. This will be handled best by retaining the generic and universal tone that has contributed so heavily to ISO 9001's success.

To help manage its portfolio of work, ISO/TC 176 developed a list of sector-specific standards entitled "ISO/TC 176 N881 Sector Specific Documentation List." It can be found at *www.tc176.org*. Figure 4.1 contains a partial listing of the ISO 9001-based sector standards.

Figure 4.1 ISO 9001-Based Sector Standards		
Industry/sector	Document identification	Document title
Aerospace	AS9100	Quality management systems— Aerospace— Requirements
Agriculture	ISO 22006	Guidelines on the application of ISO 9001:2000 for crop production
Automotive	ISO/TS 16949	Quality management systems— Particular requirements for the application of ISO 9001:2000 for automotive production and relevant service organizations
Computer software	ISO/IEC 90003	Software engineering—Guidelines for the application of ISO 9001:2000 to computer software
Education	ISO IWA 2	Quality management systems— Guidelines for the application of ISO 9001:2000 in education
Government	ISO IWA 4	Quality management systems— Guidelines for the application of ISO 9001:2000 in local government

Medical devices	ISO 13485	Medical devices—Quality management systems—Requirements for regulatory purposes
Petrochemical	ISO/TS 29001	Petroleum, petrochemical, and natural gas industries—Sector-specific quality management systems—Requirements for product and service supply organizations
Small business	HB 90.1	ISO 9000 for small business
Telecommuni-cations	TL 9000	Quality management systems—Telecommunications
Test lab	ISO 17025	Test lab accreditation, generally including instrumentation calibration

No matter how you look at it or which industries you serve, it's widely accepted that the ISO 9000 family of standards has been a solid fundamental baseline for QMS. The positive effect that ISO 9001 and the accreditation process has had on industry over the years has been the prime driver in the development of other ISO and sector-specific standards.

As the number of sector-specific standards increases, ISO/TC 176 has been careful to maintain ISO 9001's primary objective. Although minor wording differences may not ruffle any feathers in the user community, changes in the language of such an important base document can cause some serious concern.

It became obvious early in the ISO 9001 update process that developing a liaison relationship with the owners of the sector-specific documents was crucial. The owners of these documents typically have representation on various task groups while the drafting process takes place. Their liaison status with ISO allows them to comment as well as participate in meetings where issues related to the ISO 9000 family of standards are discussed. The feedback and relationship with all of the affected economic sectors helps to keep the standard credible and relevant in the marketplace.

Remember that some of these standards have incorporated the text of ISO 9001 or used its clause structure, so it was imperative that sector-specific voices be heard. This was not a frivolous concern. The volume or gravity of changes could have had serious consequences, requiring many of the sector-specific standards to be revised. It would have been irresponsible to minimize the effect that large-scale changes would have: to alienate users whose good will is essential to this process.

To maintain sustainability and value for the users of multiple standards, those responsible for generating them must pay attention to the marketplace. The technical experts of ISO/TC 176 have been vigilant to ensure that the necessary inputs are solicited from users of sector-specific standards. At the same time, they need to ensure that the outcome doesn't skew the document toward any one particular industry at the cost of another, avoid inserting conflicting language that causes problems with requirements from another industry, and prevent introducing specificity about any one of the sectors requirements, which would jeopardize the generic hallmark of ISO 9001.

ISO/TS 16949

With more than 20,000 certificates issued, there is no doubt that ISO/TS 16949 is one of the most recognizable of the sector-specific standards that are based upon ISO 9001. Due to the sheer number of these certificates, it would be remiss to not spend a little extra time discussing it. It's useful because it illustrates the same themes and challenges that undoubtedly affect the other major ISO 9001-based standards.

In lieu of providing a history lesson about how the automotive standards came into existence, this chapter highlights how ISO/TS 16949 influenced the ISO 9001:2008 amendment.

The International Automotive Task Force (IATF), which publishes ISO/TS 16949, has long recognized that ISO 9001 is a sound set of generic requirements. However, the requirements within ISO/TS 16949 give some specificity to items of importance to the subscribing members of the IATF (which are eight of the primary global automotive manufacturers).

It's somewhat ironic that the creation of ISO/TS 16949 was largely to provide supplementary requirements to ISO 9001 specific to the automotive industry, considering that a significant feature of the registration process is the addition of customer-specific requirements. This recognizes that not only are there issues of concern within the automotive industry that may need specificity, but there are also issues that are of unique concern to specific customers. This layering of requirements that make up the fabric of the ISO/TS 16949 certification process is one of the key differentiators and points of control that the IATF has placed on suppliers.

It has been said that the perceived weaknesses in ISO 9001 certification are not in the requirements document. Rather, the weakness is in the credibility of the certification process. The IATF has taken steps that appear to mitigate this issue by maintaining tight control over the process. The IATF establishes a contract with the certification bodies that perform assessments on its behalf. It has published a document titled, "Automotive certification scheme for ISO/TS 16949:2002, Rules for achieving IATF recognition," in an effort to provide fairly strict guidance for the conduct of assessments. Coupled with training requirements for assessors and witnessed audits by IATF representatives, the IATF keeps a close control of the certification process.

The liaison status of automotive industry representatives allows them to have an influence on the update to ISO 9001. In the earlier revisions of ISO 9001, representatives did not actively participate in the drafting process, although ISO/TS 16949 and the various predecessor documents, such as QS-9000, were based upon it.

The active participation of the IATF yielded some interesting discussions during the drafting of ISO 9001:2008. It has a unique perspective because the automotive industry represents a large voting block of certifications. With its tight control on certification bodies, it receives massive amounts of feedback and data. Consequently, on more than one occasion, it has influenced some of the wording and scope of ISO 9001:2008 by bringing this wealth of feedback from the user community. It has also consistently provided a good challenge to the technical experts to consider not only substantive changes, but also to balance the perception of change vs. the value of the clarifications in ISO 9001:2008.

The automotive industry has already taken the first steps to begin revising ISO/TS 16949. At a recent ISO/TC 176 meeting, automotive industry representatives explained their plans for amending ISO/TS 16949. This plan was documented in a formal resolution approved by the participating countries. This resolution states the following:

- The work underway in subcommittee 2 with imminent publication of ISO 9001:2008 includes no new requirements.

- The automotive sector document ISO/TS 16949:2002 makes complete use of ISO 9001:2000.

- ISO/TC 176 requests the IATF make the necessary updates to ISO/TS 16949. This would include incorporating the ISO 9001:2008 text and limiting amendments in the ISO/TS 16949 text to those of an editorial nature. The updated version will be balloted by ISO/TC 176.

What does this mean to users of ISO/TS 16949? Just as with ISO 9001:2008, there will be little or no effect to most users. It's also anticipated that the automotive industry will follow a similar transition path as ISO 9001:2008, giving both certifications equal value.

CHAPTER 5

THE ISO STANDARDS-DEVELOPMENT PROCESS

A s a user of ISO 9001, you may wonder how its 2008 amendment came to be and what the process is for developing standards. Before considering the drafting process, the first step is to understand the rules and infrastructure within which ISO 9001 was developed.

In this chapter we'll examine three phases of the process required to complete ISO 9001:2008: justification for the amendment, the design specification that guided the drafting, and the drafting process itself.

JUSTIFICATION FOR AMENDING ISO 9001

Management system standards are administered by technical committees that use ISO Guide 72 to develop the standards. For the ISO 9000 family of standards, this technical committee is ISO/TC 176. ISO Guide 72 requires ISO/TC 176 to conduct a systematic review of ISO 9001 every five years. Some believe that the actual revision must take place within the five-year time period, but this isn't the case. If it was, ISO/TC 176 would have to immediately start the next revision after each amendment is published to meet the five-year timetable.

The results of the systematic review will yield one of several decisions: withdraw, revise/amend, confirm, or confirm with error corrections. In 1993 the conclusion reached from the systematic review of ISO 9001 and ISO 9004 was to make a minor amendment to ISO 9001 and to revise ISO 9004.

The member bodies of ISO/TC 176 indicated that they were satisfied with ISO 9001:2000 and that there was no need for substantial revision. So why wasn't the standard simply confirmed again if the participating countries realized that there was only a need for minimal changes?

One of the main reasons is timing. ISO 9001 and ISO 14001 have been on different revision cycles, making it difficult for their users and the technical experts who created them to maintain compatible requirements. Taking this compatibility to the next level is definitely connected to the ability to edit these documents at the same time.

Because the update to ISO 14001 was published in 2004, it wasn't going to be ready for a systematic review for another five years. ISO 9001 was last amended in 2000. Considering how many things can change over the course of several years, waiting until 2009 to conduct the next revision could call into question ISO 9001's continued relevance. This is why many countries voted to amend the document rather than confirm it.

After the review was conducted, ISO Guide 72 required the completion of a justification study. For the purposes of ISO 9001 and ISO 9004, this study was completed by the working group (WG) project leader and the ISO/TC 176 subcommittee (SC) 2 secretariat. They incorporated a review of the market as well as feedback provided by participating countries in the systematic review.

DESIGN SPECIFICATION FOR THE AMENDMENT

After the decision was made to amend ISO 9001, ISO/TC 176 developed a design specification. The purpose of the design specification was to provide a framework to meet the needs of users, improve known issues, and ensure that the changes did not go beyond amendment-level.

The design specification was circulated to the committee's voting members for approval. The voting process includes the opportunity for members to make comments that must be considered before an approval is finalized.

The design specification serves multiple purposes. It defines the scope of the amendment project, keeps ISO/TC 176 from deviating from the approved project specification, and is utilized by the task group (TG) responsible for validating the final product to ensure that the output of the drafters' work matches the defined inputs.

Because there is no formal definition of an amendment, special attention was needed to clarify what types of changes fell into the category of amendment. The types of changes identified in the design specification included those that would address such concerns as clarification, compatibility with ISO 14001, and improved translatability. However, it wasn't possible to address these types of potential changes without determining if there was an effect on users. For this reason ISO/TC 176 developed a risk matrix to weigh the benefit of any change against the changes (or even perception of changes) users might need to make to their quality management systems (QMS). Each proposed change was analyzed with this matrix and only the items that were high benefit and low impact could be considered.

The risk matrix was successful in controlling the extent of the changes to ISO 9001. Changes that didn't meet the criteria were set aside to be considered in the next revision of the standard.

After the matrix was created, the task groups began the actual amendment work. To meet user needs, the editing focused on published sources of information. The inputs that were considered included:

- ISO/TC 176 sanctioned interpretations
- ISO/TC 176 SC 2/N 683R, ISO/TC 176/WG Interpretations—Collation of interpretations and inputs to SC 2
- Systematic reviews conducted on ISO 9001 and ISO 9004 (ISO/TC 176 SC 2 documents: N666, N667, N676R, N677R)
- Web-based feedback survey on ISO 9001 and ISO 9004 conducted by ISO/TC 176 SC 2, with the assistance of ISO Central Secretariat (ISO/TC 176 SC 2 documents: N631, N668, N681, N705)
- National feedback surveys and research, including those from the United

States, Japan, Germany, and Canada (ISO/TC 176 SC 2 documents: N607–1, N607–2, N637, N680)

■ ISO/TC 176 SC 2/N682, *Justification study for an amendment to ISO 9001:2000 and a revision to ISO 9004:2000*

■ ISO 14001:2004

■ ISO Guide 72

■ The ISO 9001:2000 Introduction and Support Package set of documents

■ ISO/TC 176 SC 2/N307—Design specification for ISO 9001:2000

Let's discuss a few of the inputs.

TECHNICAL EXPERTS

Who are these technical experts and how are they assigned? If you were to look at the credentials from members of ISO/TC 176, you would find a wide range of backgrounds represented. The members represent industry, academic institutions, and standards bodies. They might be CEOs, high-ranking officials of a corporation, or have doctorates in their particular field of expertise. They are often seen as the best in their fields, and, most important, they check their egos at the door.

Since the last revision to ISO 9001, the technical experts who write the standards have learned new techniques, been exposed to new markets, and generally expanded their comprehension of quality management. Some retired and new candidates replaced them, bringing along their own knowledge and unique perspectives. The continued exchange of ideas contributes to the integrity of the standards they ultimately produce.

When staffing a task group such as the one that worked on the amendment to ISO 9001, the secretariat asks each country and approved liaison group to nominate potential members. To ensure balance, each country and liaison, regardless of its size and influence, is given one or two positions on the task group. With a document such as ISO 9001, each country and group wants to be represented on the task force, so this results in a very large group. Task groups are easiest to manage when there are twelve to twenty members. In the case of this amendment, there were more than fifty active members.

Because of the critical nature of ISO 9001, it's rare that a technical expert is nominated to participate on an ISO 9001 task group without past experience on another standard in the ISO 9000 family or support standards. It's important that you be recognized by your peers on the task group as an expert to efficiently negotiate your positions.

It's also important to remember that once you're assigned to the task group, you are not representing your country but serving as a neutral technical expert. This doesn't mean that members don't try to look after the best interests of their country, and at times it's important that they elaborate on a comment from their country or liaison group, but for the most part the experts remain neutral.

Using its members' collective body of knowledge, the task group works to create positive change for all user groups and industries. These technical experts understand their roles: to ensure that they produce a document that meets the design specification while meeting user needs. It's not always easy. The debates this group had in developing consensus positions were arduous, but they were conducted with respect and the understanding that even heated dialogue helps produce broad consensus positions.

GLOBAL USER QUESTIONNAIRE/SURVEY

In 2004, ISO/TC 176 conducted a global survey pertaining to everything from organizations' demographics to specific questions regarding ISO 9001 clauses. Sixty-three countries submitted responses from all four of the product categories (hardware, software, processed materials, and services).

Out of the responses received, 80 percent of the respondents were satisfied with the 2000 version of the standard. They provided more than 1,400 comments, all of which were considered during the editing process. Some of the comments were single thoughts on particular issues. In some cases, ISO/TC 176 determined that an issue shouldn't be changed without an obvious trend in the feedback. In addition, some of the feedback would have required a change that would extend beyond the scope of an amendment.

Some of the countries represented on the committee had conducted independent studies of ISO 9001:2000 at the time of the transition. Their feedback was complementary to that of the global surveys and provided country-specific feedback, and it was submitted for consideration in the amendment. Specifically, the

United States, Japan, Canada, and Germany submitted their feedback and lessons learned.

Most of the comments related to one of the following seven clauses:

- 4.1, Quality management systems—general requirements
- 4.2, Quality management systems—documentation
- 6.2, Human resources
- 7.3, Design and development
- 7.5, Production and service provision
- 8.2, Monitoring and measurement
- 8.5, Improvement

MARKET JUSTIFICATION STUDY

ISO/TC 176 also considered the results of the market justification study it performed. The report included information regarding relevant issues in the marketplace, such as compatibility with other management system standards (e.g., ISO 14001) and the identification of interpretation issues that required clarification. The justification study also identified the worldwide feedback survey and the needs of sectors in determining the need for an amendment.

INTERPRETATIONS PROCESS

The sanctioned interpretations submitted by the working group on interpretations were considered during the editing process. To ensure that the intent of the interpretations was maintained, the working group on interpretations provided a liaison member to participate in the editing group.

ISO/TC 176 also considered those interpretation issues that hadn't been resolved. In reality, examining these issues had more potential benefit because an interpretation could not be reached on the proposed issue. ISO/TC 176 strove to resolve these issues without changing requirements.

ISO 14001

Part of the design specification that guided the work of ISO/TC 176 was to ensure ISO 9001:2008's continued and enhanced compatibility with ISO 14001, where it was practical. Indeed, there were many conversations held during the drafting process to review the language to see if it was appropriate to use the same wording. You may find as you read the amendment that there were cases in which changes were made specifically with this in mind.

However, if you're looking for identical wording in all possible cases, you may be disappointed. ISO/TC 176 determined that there were instances where adopting the language that was used in ISO 14001 would have created too large of a departure from the current language, resulting in either the introduction of a change in the requirement or the perception that a requirement had changed—both scenarios being contradictory to the intent of ISO/TC 176 and the design specification. This may seem like a fine point and more "standards speak," but the technical experts were very conscious to weigh the perception of change in the user community against the benefit of added clarity or compatibility.

DRAFTING THE AMENDMENT

As with most new or revised standards, the first version (or two) is referred to as a working draft (WD). Feedback at this stage on the drafting of the document is limited to the technical experts within the WG, as WDs are generally not circulated outside of the WG. The WG is the cumulative umbrella for all the task groups (TG) working on the project.

Work at the WD level involves meetings, teleconferences, and hundreds of e-mails. Decisions are achieved through a consensus process. Considering that we're dealing with colleagues from dozens of countries around the globe representing a myriad of cultures, languages, and industries, this process is nothing short of extraordinary. The price that is paid in perceived delays is easily outweighed by the value of this inclusive model.

At the WD level, only the technical experts provide comments. When the WD reaches adequate maturity (as determined by the TG with input from other WG members), it is released as a committee draft (CD). There can be multiple CDs depending on feedback from the voting members and the user community.

STANDARDS-DEVELOPMENT PROCESS

CDs may be issued for comment only or for ballot. For the purpose of this amendment, ISO 9001 was originally distributed as CD for comment only.

During the CD for comment stage, participating countries also distribute the document for review by technical experts of their mirror committees (in the

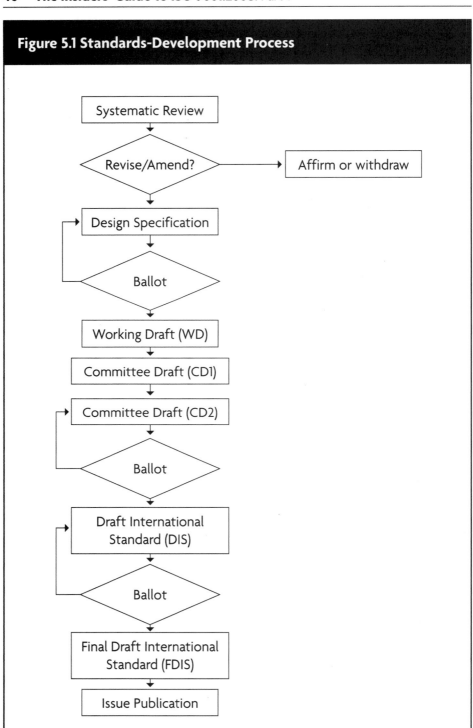

Figure 5.1 Standards-Development Process

United States, it is the U.S. Technical Advisory Group to ISO/TC 176). The document is also available to the general public for comment.

Each country develops a consensus position on its comments and submits it to the secretariat of ISO/TC 176 SC 2 (the subcommittee that is responsible for the content of ISO 9001) for consideration.

A standard undergoing change can also be distributed as a CD for voting. This means that, in addition to providing comments, each participating country also votes on whether it approves the document.

For the purpose of ISO 9001:2008, its CD was determined to be mature enough after it was circulated for comments, so a CD for ballot was determined to be unnecessary. This decision was made by resolution and a vote at an ISO/TC 176 SC 2 meeting.

By seeking a ballot as well as comments at the CD stage, the technical experts were able to eliminate six to nine months from the original publication schedule; ISO 9001:2008 was originally scheduled to be published in the spring of 2009.

Subsequent to the CD stage, the document went through a succession of balloting and review of comments. After the CD phase, the document was published as a draft international standard (DIS) and was released for public comment. These were collected and submitted for review by the member bodies of ISO/TC 176.

After another round of drafting based on the feedback from the comments and voting, the document progressed to the final draft international standard (FDIS) phase. At this point it was virtually ready for publication. The member bodies of ISO/TC 176 were balloted for final approval of the document and it was prepared for final release through the editorial process within the Central Secretariat of ISO.

During the amendment drafting process, face-to-face meetings of ISO/TC 176 members were conducted to review comments. Each comment was reviewed and dispositioned based on its acceptance or nonacceptance for editorial or technical reasons. During this review, some comments were quickly accepted because they met the intent of the standard and were related to areas identified in the design specification as targets for clarification. For example, the notes added to clause 4.1 to clarify what is meant by outsourcing were quickly accepted. However, adding words that improved the clarity without confusing users or giving the perception of substantial change was more difficult. For that reason, the technical experts spent a lot of time balancing improvement with the perception of change. If the technical experts could not agree that the change did not improve clarity without risking confusion or misperception, the primary position taken was to leave the ISO 9001:2000 text unchanged.

In some cases, the committee identified items that would have improved ISO 9001, but the changes weren't made because they went beyond the scope of the amendment. This included a proposal on how required records are identified. Any changes or comments that went beyond the scope of the amendment by either adding or deleting requirements or changing documentation requirements were also tabled for the next revision. This included concepts such as risk management or structural changes resulting from relocating requirements within the standard.

The interpretations were considered as input, but there were some changes that couldn't be made. At times, changes made to the standard to eliminate the need for an interpretation could have changed requirements. For example, the phrase "in a form" has begged the question of whether a form is required to provide the output. However, changing this wording implied a change in requirement. The technical experts also erred on the side of caution when making grammatical changes. By making extensive changes to the standard's text, users could perceive that the requirements had been changed instead of making an improvement to language.

In addition, the committee made changes only to improve compatibility with ISO 14001 where the change did not add or reduce existing ISO 9001 requirements. It's anticipated that the next revisions of ISO 9001 and ISO 14001 will be conducted concurrently. This means that there will be greater opportunity in the future for improved compatibility.

CHAPTER 6

COMPATIBILITY WITH ISO 14001

Many companies have already chosen to establish management systems compliant to both ISO 9001 and ISO 14001. Consolidation of quality and environmental management systems includes aspects such as documentation, management reviews, internal audits, and corrective and preventive action systems. Organizations that use ISO 9001 and ISO 14001 will tell you that there are many opportunities for synergy in synchronized standardization approaches and documentation and execution strategies.

Recently, the technical committee that has ownership of ISO 14001 (ISO/TC 207) reported that it received many comments during its review of ISO 14001 indicating a broadly shared expectation that the standard's revision will be coordinated and aligned with a future revision to ISO 9001. The concept of alignment between these standards has been the subject of much discussion during many meetings. Quite some time ago, the technical management board

(TMB) of ISO established teams to work with ISO/TC 207 and ISO/TC 176 to develop a joint vision for what this alignment may look like. The task was to ensure that all future revisions to the two requirements documents would not only have common text and terminology wherever practical, but that they would also have common elements.

During the ISO 9001 amendment process there was a concerted effort to increase compatibility and alignment with ISO 14001 wherever possible and practical. In particular, the drafters considered common text and terminology. There was, however, a delicate balance between the desire to improve compatibility and introducing changes just for the sake of alignment, especially where that could create confusion or misunderstanding for the users of ISO 9001. The committee agreed to avoid making changes to increase compatibility without any offsetting benefit to clarification.

To that end, the team employed a guideline that when there was a specific input (such as those discussed earlier) requesting clarification, the text and terminology within ISO 14001 would be consulted as a primary source documents. In fact, during drafting meetings, there was typically a person designated to review ISO 14001 for recommendations.

As we move into the later section of the book and use matrices to illustrate the changes incorporated into ISO 9001:2008, you will see some areas in which the changes were at least partially justified by increased compatibility. In particular, some of the text and terminology from ISO 14001 was incorporated into ISO 9001 clauses, such as 4.2.4, Control of records. The wording and even the structure of the clause were adjusted to increase compatibility.

Although it can be said that it was always considered, it was not practical for the technical experts to adopt consistent wording in every case. There were times when language was used from ISO 14001 in the early draft stages that resulted in many comments from users and member nations reviewing the document. The changes not only didn't add clarity to the amendment, but they also made it harder to understand.

As the conversation inevitably switches from the ISO 9001:2008 amendment to future revisions, there is a clear expectation that there will be joint work and coordination between the drafters of environmental management system standards and quality management system standards. Considering how difficult it has been to get rooms full of quality experts to come to consensus, it will be very interesting to see how the consensus process survives the additional input and complexity that more experts with different perspectives and needs will bring.

CHAPTER 7

UPDATE ON ISO 9004

The revision to ISO 9001:2000 has held center stage in the standards arena for the last three years. Considering the number of organizations holding ISO 9001 certificates, the attention is not unwarranted.

The work being done to revise ISO 9004:2000 has received significantly less attention. However, given the fact that it's not nearly as popular as ISO 9001, the amount of interest it has generated is great news. It's been noted that part of the reason ISO 9004 is less popular than its stellar sister ISO 9001 is that it's a guidance document. It's not possible to be certified to ISO 9004. It was never intended to be a conformance document but that doesn't diminish the value that can be derived from its use.

The present version has been particularly useful in several sectors, most notably as a guide for European medicines agencies and as the architecture for IWA2:2007, Quality management systems—Guidelines for application of ISO

9001:2000 in education. Input from surveys and other market inputs indicate that many organizations have used ISO 9004 to improve how they do business.

The title of the existing version of ISO 9004:2000 is Quality management systems—Guidelines for performance improvements. The proposed title for ISO 9004:2009 is Managing for sustained success of an organization—A quality management approach.

ISO 9004:2000 was conceived as a guideline that would facilitate organizations' ability to use their quality management systems (QMS) to improve. These improvements relate to a myriad of features, including product, process, internal culture, and, ultimately, the bottom line. The document was not intended to be a how-to guide for ISO 9001 conformance, although it has been employed by some for that purpose. For everyone else, it was a great resource for moving from simple conformance to having a QMS that created value. It created more worth than the achievement of a certificate.

The document was based on the endorsement of the following eight quality management principles:

- Customer focus
- Leadership
- Involvement of people
- Process approach
- System approach to management
- Continual improvement
- Factual approach to decision making
- Mutually beneficial supplier relationships

These principles are not requirements. Rather, the expectation is that a well-developed and effective QMS will manifest these principles. Therefore, to experience improvement and the corresponding benefits, a commitment to these quality management principles would be anticipated.

The use of the quality management principles as a foundational premise has been retained in the new version of ISO 9004. In terms of structure and appearance, ISO 9004:2009 will be markedly different. Unlike ISO 9001, ISO 9004 is experiencing a complete overhaul. Although the essential concept of improvement has also been retained, the manner in which it is presented is much more focused. The technical experts who are developing ISO 9004:2009 realize that it isn't enough to talk about improvement. ISO 9004:2009 must have more specificity regarding the how and why. The technical experts also believe that it's important to introduce the concept of sustaining the organization's success over time.

Three concepts drove the revision:

- Meeting the needs of interested parties
- Strategic planning
- Self-assessment

MEETING THE NEEDS OF INTERESTED PARTIES

ISO 9000:2005 defines an interested party as: "person or group having an interest in the performance and success of an organization." Examples of interested parties include customers, financial institutions, suppliers, society, and the people in the organization. For an organization to succeed over time, it's necessary for it to monitor and assess the needs and expectations of these parties and the risks they engender for the organization. The chart shown in figure 7.1 illustrates some of the typical connections between interested parties and the concerns the organization should address.

Figure 7.1 Interested Parties and Organizational Concerns	
Customers	Increase or decrease in requirements New designs requiring acquisition of new technology Increase in competitors selling to organization's customers Demands for faster delivery
Suppliers	Relocation offshore Decreased ability to meet organization's requirements Loss of certification Global raw material shortages

Financial institutions	Tightening of lending policies Availability of funds to lend Changes in rules and practices
Society	Environmental concerns Changes at the local community level Revisions to statutory and regulatory requirements
Internal personnel	Availability of competent staff Need for fair wages Opportunity for learning and advancement Safe work environment

The concept of interested parties is pivotal to sustaining success. Failure to appropriately address their concerns can have dire consequences for an organization.

The success of organizations isn't just an altruistic goal. In an increasingly global economy, the failure of one organization represents a breakdown in the supply chain of another organization. Having suppliers certified to ISO 9001 is great, but knowing that they have systems and processes to ensure continued viability and success is significantly more crucial.

STRATEGIC PLANNING

ISO 9004 seeks to improve the manner in which the role of management in the QMS is related to organizational planning. Specifically, it defines the planning activities as a process dependent on the consistent analysis of reliable data derived from an understanding of the needs of interested parties and the monitoring of performance indicators. The other input into strategic planning comes from the third factor that drove this revision—self-assessment.

SELF-ASSESSMENT

ISO 9004 in its current incarnation, as well as its 2009 revision, endorses the need for organizations to periodically assess their status in relation to a variety of indicators. The indicators are aligned to the new document structure but are still reflective of the eight quality management principles. They look at an organization's maturity based on achievement of general benchmarks in practices. For example, an organization's practices vis-à-vis communication might range through several levels from reactive and sporadic to proactive, planned, and consistent. The output of the self-assessment helps the organization to identify areas warranting attention and improvement.

Other concepts received greater elaboration from the previous version, particularly the identification and management of resources and the concept of innovation.

The structure of ISO 9004 has been completely revamped. ISO 9004:2000 was structured identically to ISO 9001 and included the entire text of ISO 9001. This gave rise to the expression, "the consistent pair." The need for consistency between the two documents is irrefutable. However, going forward that consistency will not be so transparent.

The technical experts decided that it would be possible to retain the compatibility between the documents without explicitly referencing ISO 9001 in each section. By refraining from this explicit reference it became possible to make the document relevant to users of other management system standards. This would serve a broader range of industries and increase the number of organizations that could derive benefit from ISO 9004. It was decided that, although compatibility with ISO 9001 was important, the opportunity to reach a wider audience (including those who did not necessarily conform to ISO 9001) was desirable.

There was perhaps only one unfortunate outcome to the differing paths ISO 9001 and ISO 9004 took for these revisions. ISO 9001, being an amendment, required less time. It made no sense to delay its release. ISO 9004 has been so completely altered that it looks like a new document. The changes took proportionately longer and so it will not be ready for release until 2009. This has resulted in the uncoupling of the consistent pair. They won't be published at the same time. This is only detrimental to those individuals and organizations that had sought to use it as a how-to manual—a purpose for which it was not intended.

As of the publication of this book, the proposed structure of ISO 9004:2009 is

Figure 7.2 Proposed Structure of ISO 9004:2009

CHAPTER 8

LOOKING AHEAD

I SO 9001 has been around for more than twenty years. It has criss-crossed the globe, expanding the burgeoning ranks of users and penetrating markets, industries, and professions in ways that no one could have anticipated. This broad appeal is largely due to its elegant minimalism and generic applicability. Its enduring success relies on its continued relevance to the user community. In essence, the technical experts must pay attention to their customers.

This perception has allowed the technical experts to take the prudent approach of issuing an amendment with the understanding that a more extensive revision would be needed in the next decade. As we've already seen, it takes a while to bring our revised product to market. Revisions to ISO 9001:2000 were initiated at the end of the last century. By the time we begin work on the next revision, the current requirements will have been in place for over a decade.

A lot can happen over the course of a decade. Consider some of the pivotal experiences of the last few years. Y2K opened our eyes to the vulnerability inherent in software interfaces. Outsourcing and offshoring has increased exponentially. The risks associated with reliance on remote suppliers have been magnified. The Internet now permeates all aspects of commerce. There is an ever-growing need for international standardization to ensure safety, reliability, and consistency—not only in products, but also in the processes that enable organizations to bring products to market. These are just some of the issues that will drive the next revision of ISO 9001.

There will be four categories of inputs in the next revision of ISO 9001. The first input we've just mentioned: all of the technological and global issues that affect the market. The second input will be from all the great ideas that were mentioned in earlier chapters that came out of ISO 9001:2008 but could not be used because they exceeded the scope of the design specification. The third input will be from the user community through surveys and other methods that will be developed to aggregate feedback. If the standard is to endure, we must apply the same requirements for monitoring, focus on the customer, and continual improvement that are hallmarks of the standard itself. The fourth and final input will relate to the needs and, conversely, influence of sector-specific standards.

TECHNOLOGICAL AND GLOBAL ISSUES

Our reliance on electronic media for communication, document control, marketing, training, and e-commerce has grown exponentially during the last fifteen years. Internally, organizations rely heavily on software to manage documentation, notification, scheduling, project planning, product design, traceability, and general communication.

ISO 9001 contains language addressing all of the above in pre-electronic era terminology. With the exception of the occasional note relating to electronic media, the standard doesn't reflect the explosion of technology or its effect on our lives. The field of information management technologies is growing. Because this factor permeates every aspect of the modern organization, it will be better addressed with the due diligence that accompanies the next major revision.

OUTSOURCING AND OFFSHORING

The Internet has facilitated outsourcing and globalization of the supply chain. Within the context of ISO 9001, outsourcing refers to the processes needed by the organization that are performed by an external entity. The decision to outsource can be based either on the determination that the organization does not have the requisite resources or ability to effectively implement the process, or that it would be financially advantageous to have someone else do it. Typical examples include third-party testing and verification, development of user manuals, packaging, heat-treating, and technical support.

ISO/TC 176 has recognized that outsourcing is a growing aspect of many organizations. It serves to extend product offerings, increase after-market service opportunities, and augment capabilities. Outsourcing can reduce overhead, defer the need for capital expenditures, mitigate risks resulting from spikes in demand, and contribute to lean initiatives. Currently, ISO 9001 has limited language in section four, addressing the requirements surrounding outsourced processes. As outsourcing becomes more widespread, it will be appropriate to insert additional language proportionate to the prevalence of the practice. Elaboration of text appropriate to the prominence of outsourcing practices would be an enhancement that would serve the user community.

A growing amount of outsourcing is going offshore, as have many other links in the supply chain. Raw materials, components, finished goods, and customer service now come to us from almost every corner of the globe. The authors of ISO 9001 anticipate that the standard will inevitably need to more extensively address requirements unique to these long-distance suppliers. There are specific constraints that accompany global supplier relations: distance, transportation, time zones, cultural differences, communication, and infrastructure reliability can all affect intercontinental partnerships. Although ISO 9001 addresses supplier-related issues, the text will need to be revised to remain relevant to evolving market practices.

GREAT IDEAS THAT AROSE DURING THE REVISION PROCESS

It's a little daunting to think that before the ink is dry on ISO 9001:2008, the steps begin for the next revision cycle. You have learned that due to the scope of an amendment many of the good ideas submitted through the commenting process or evaluated from the interpretations process could not be incorporated into ISO 9001:2008. That doesn't mean they will be forgotten. However, it cannot be assumed that just because an idea was considered during this amendment process that it will be automatically included in the next revision.

However, steps are being taken to capture these ideas. The leaders of the drafting committee for ISO 9001:2008 will prepare a report of their activities. This report will include how the drafting was completed as well as lessons learned and best practices established. It will provide ISO/TC 176 with the information it needs to improve the next revision.

In addition, during the creation of this report, every comment that was set aside during the review period for ISO 9001:2008 will be considered in aggregate. Like concepts will be grouped together and included in an input document for the next revision. It's anticipated that this input document will include topics such as risk management, recommendations on how the need for records should be identified in the standard, and overall structural changes to the standard that could improve compatibility between ISO 9001 and ISO 14001 or understanding by users. Other comments considered to have merit—regardless of whether they indicate a particular trend—will also be included.

It can also be assumed that sector-specific documents will be used as one of the inputs for the next revision. Ideas that can be incorporated into ISO 9001 which can reduce the need for sector-specific versions of the document should be considered. However, keep in mind that ISO 9001:2008 focuses on being a generic standard that can be applied by any organization regardless of its size or product type. Including too much text from these sector-specific documents or including other concepts could reduce this focus.

SURVEYS/FEEDBACK FROM USERS

Once ISO 9001:2008 launches, mechanisms will be established to gather data from the user community. This information will be considered from the onset of the revision process and will be invaluable to the development of its design specification.

SECTOR-SPECIFIC STANDARDS

Over the years the proliferation of sector-specific standards has contined unabated. In the last two years, standards have been added relating to crop production and the petroleum industry. They join the now-mature standards previously mentioned in such areas as medical devices, automotive, telecommunications, and aerospace. The standards field has been further augmented by standards relating to environmental issues, health and safety, and laboratory accreditation.

The various industry bodies sponsoring the sector-specific standards are also rightly concerned that a major change to ISO 9001 will inevitably necessitate revisions to their unique standards if they wish to retain compatibility with ISO 9001. In addition, some of the standards, most notably the automotive industry's ISO/TS 16949, rely on ISO 9001 as the recognized QMS models for which the second- and third-tier suppliers for which registration to standards such as ISO/TS 16949 would be inappropriate or unnecessarily burdensome. Changes, therefore, will not be made lightly or frivolously.

Conversely, changes in the specific industries and, in some cases, in the regulatory industries that provide oversight, may also influence the language and requirements of the next revision.

ISO 14001 and OHSAS 18001 differ somewhat from the other sector-specific standards. They relate respectively to environmental issues and health and safety concerns. ISO/TC 207 is responsible for the ISO 14000 series of standards. For several years, ISO/TC 207 and ISO/TC 176 have striven to align the language of ISO 9001 and ISO 14001. This is expected to be another of the myriad inputs into the next major revision of ISO 9001.

OHSAS 18001 is a newer arrival and, although a formal commitment to alignment has not been made, the authors of ISO 9001 will not be able to ignore its growing influence.

SARBANES-OXLEY

The Sarbanes-Oxley Act of 2002 has inadvertently created a new market of users. Due to the requirements around record retention, document control, and the processes that eventuate control of both, organizations are looking to ISO 9001 for guidance on managing these aspects of their organizations. Although ISO 9001 doesn't encompass requirements for accounting and finance, we cannot ignore this new group of users who are finding value in this QMS standard.

CAPABILITY MATURITY MODEL INTEGRATION (CMMI)

Over the years, the Software Engineering Institute, with input from government and industry, has developed a model for addressing process improvement. Capability maturity model integration (CMMI) is most often associated with the software industry. Experts utilizing both CMMI and ISO 9001:2000 have recognized that there is considerable compatibility between the bodies of knowledge encapsulated in each model. It's apparent that there is an opportunity for mutual learning and increased understanding between users of either model. Bearing in mind the earlier comments concerning emerging technologies and our increased reliance on electronic media, including the discussion of the benefits of CMMI, during the next revision of ISO 9001 could only serve to strengthen the document.

INTEGRATED MANAGEMENT SYSTEMS STANDARDS

There is additional work underway that may bring even more challenges to the drafting process. The ISO technical management board has commissioned additional work to expand the dialogue for considering a joint vision to encompass *all* management system standards. What are the potential inclusions in the joint vision along with ISO 9001 and ISO 14001?

- Risk management
- Health and safety
- Next management system standard (MSS)
- Others?

How will this all progress? No one is sure. One of the greatest assets of ISO 9001 is the fact that it can be applied to any company of any size within any industry. However, one of the greatest criticisms is the lack of specificity that this flexibility brings. It's very difficult, however, for many to see how a document would be crafted that can be general enough to fit the potential requirements for all management system standards while still being specific enough to be useful.

Many of these factors were barely perceptible when decisions concerning ISO 9001:2008 were made by ISO/TC 176. The amendment provides a well-controlled intercession between the 2000 release and the next major revision. During the intervening years, the technical experts will have the time to solicit more input from the user community, aggregate and analyze the data, and develop the next generation of the ISO 9001 standard to ensure its continued relevance and success.

Ultimately, the future of ISO 9001 relies on two things: retention of the features that have contributed perennially to its popularity, and periodic revisions that ensure its continued applicability in our ever-changing world.

As we dive into the next section and explore the nature of the amendment's wording, it will be evident that there will be no need for major changes to your systems, documentation, or processes at this time. Major changes are on the horizon for a future day.

PART 2

CLAUSE-BY-CLAUSE REVIEW

The following templates will guide you through an assessment of your ISO 9001:2000-based quality management system (QMS). The amendment provides organizations an opportunity to take a look at their QMS through fresh eyes, providing them with the opportunity to experience increased learning and awareness of the status of various aspects of the QMS.

The assessment can be likened to a gap analysis wherein you assess four factors for each clause or subclause. The questions to be answered for each are:

- Is the requirement adequately documented?
- Is the requirement correctly implemented?
- Does the manner of implementation reflect the intent of the standard?
- Is the requirement, as documented and implemented, a benefit to the organization?

The outcome of the assessment will yield the following information.

■ Recognition of problems and omissions

■ Identification of risks due to inadequate implementation or ambiguous documentation

■ Insight that may spark new ideas for enhancement and improvement

■ Confirmation of adequacy, correctness, and completeness

Rather than looking upon the exercise as an onerous task to ensure conformance, seize the serendipitous opportunity to assess your organization's unique QMS from a slightly altered perspective.

HOW TO USE

Each of the templates includes the following information:

■ Title of the clause/subclause

■ Subject of the clause/subclause

■ Nature of change*

■ Explanation and rationale for the change

■ Guidelines on what to check for to ensure conformance.

*Note: Where there were both clarification and editorial changes, the first check box ("clarification") is the only one that is checked. It may be assumed that where there was a need for clarification, the outcome generally included editorial changes. In contrast, editorial changes may have been as insignificant as the correction or re-location of a word or phrase.

Make notations for each clause/subclause. When the assessment is complete, the consolidated notes will define the extent of changes that the organization will need to implement. For some organizations, the end result will be a brief to-do list. For others, it will be the outline for a much larger project. In either case, the templates will allow you to define the scope and the magnitude of your organization's own amendment process.

Title of clause/ subclause	Introduction 0.1 General		
Intent of the text	This is the first part of the introduction. It describes the nature, intent, and use of the document.		
Nature of the change	☑ Clarification/ correction	☐ Editorial/ consistency	☐ No change

What changed	Why the change
Added the phrase "its organizational environment, changes in that environment or risks associated with that environment" to the first paragraph.	Adding the reference to risk in the general section of the standard provided the opportunity to link in risk management. There was a strong interest in risk management in the standard. Referencing risk here brings in the aspects of risk management without changing requirements.
Revised "processes employed" to "processes it employs."	This is a grammatical change to better complement the other changes made to this paragraph.
Added "statutory and" to regulatory in the third paragraph.	Adding "statutory and" to this paragraph provides consistency within the standard and specifically with text in clause 7.2.1.
Changed "requirements for products" to "requirements applicable to the product."	Clarify that requirements that need to be considered in the quality management system are those applicable to the product.

☑ Check for conformance to requirement
N/A

Title of clause/ subclause	Introduction 0.2 Process approach		
Intent of the text	This paragraph describes the process approach, as applied to this standard and its relevance to a well-developed QMS.		
Nature of the change	☑ Clarification/ correction	❑ Editorial/ consistency	❑ No change

What changed	Why the change
Changed "identify" to "determine" in the second paragraph.	This change was made to consistently use the word "identify" throughout the standard. It was determined that "determine" better described the activities of this clause.
Changed "an activity" to "an activity or set of activities" in the second paragraph.	Clarify that a process can be one activity or multiple activities.
Added the phrase "to produce the desired outcome" to the third paragraph that explains the application of the process approach.	This change was added to the standard to emphasize that the results from building the product or providing the service do matter. Organizations that are certified to ISO 9001 must realize that in addition to obtaining certification, their output matters.

☑ Check for conformance to requirement
N/A

Title of clause/ subclause	Introduction 0.3 Relationship with ISO 9004		
Intent of the text	This paragraph describes the relationship of ISO 9001 to ISO 9004.		
Nature of the change	☑ Clarification/ correction	❑ Editorial/ consistency	❑ No change

What changed	Why the change
Revised the first paragraph to indicate that ISO 9001 and ISO 9004 have been designed to complement each other. Added a note to indicate that ISO 9004 is still undergoing revision at the time of the publication of ISO 9001.	Because ISO 9004 is still undergoing revision, the technical experts did not have enough data to clearly state the relationship of ISO 9001 and ISO 9004.
Revised the third paragraph to indicate the scope of ISO 9004.	This clause was also updated to describe that ISO 9004:2009 provides guidance to management for achieving sustained success for any organization.

☑ Check for conformance to requirement
N/A

Title of clause/ subclause	Introduction 0.4 Compatibility with other management systems		
Intent of the text	This section of the introduction discusses the alignment to ISO 14001 and relationship to other QMSs.		
Nature of the change	☑ Clarification/ correction	❑ Editorial/ consistency	❑ No change

What changed	Why the change
Revised the first paragraph to indicate that the provisions of the most recent version of ISO 14001 were taken into consideration during the revision.	Compatibility with ISO 14001 was one of the key inputs for this amendment. It is important to remember that the requirements of ISO 14001 were taken into consideration. In some instances, to change the language of ISO 9001 to more closely match ISO 14001 would have resulted in a new or reduced requirement. In these cases, the language which would keep ISO 9001 within an amendment was used.

☑ Check for conformance to requirement
N/A

Title of clause/ subclause	1.0 Scope 1.1 General		
Intent of the text	This section provides general guidelines for the scope of the document and articulates the need for the QMS to meet customer and regulatory requirements and to enhance customer satisfaction.		
Nature of the change	☑ Clarification/ correction	❑ Editorial/ consistency	❑ No change
What changed		**Why the change**	
Revised this section to include statutory requirements.		Adding "statutory and" to this paragraph provides consistency within the standard and specifically with text in clause 7.2.1.	
Revised note in ISO 9001:2000 to clarify what is meant by product. Product includes output resulting from product realization processes.		Revised note 1 to help clarify outsourcing and emphasize once again that the intended results of the process matter.	
Note 2 was added to provide an explanation that statutory and regulatory requirements can also be called legal requirements.		Clarified with a note an explanation of statutory and regulatory requirements.	

☑   Check for conformance to requirement
N/A

Title of clause/ subclause	1.0 Scope 1.2 Application		
Intent of the text	This section asserts the generic nature of the standard and its applicability to all organizations. Of particular note is the requirement to articulate any exclusions and the limitation that exclusions may only be claimed for appropriate requirements in section 7.		
Nature of the change	☑ Clarification/ correction	❏ Editorial/ consistency	❏ No change
What changed		Why the change	
Revised paragraph 3 to include statutory requirements.		Adding "statutory and" to this paragraph provides consistency within the standard and specifically with text in clause 7.2.1.	

☑ Check for conformance to requirement
Review the application of this standard to the scope of your business operations specifically in accordance with clauses 4.1 and 4.2.2.

Title of clause/ subclause	2.0 Normative reference		
Intent of the text	This section lists the relevant international standards that, through reference, constitute requirements of ISO 9001:2000.		
Nature of the change	☑ Clarification/ correction	❏ Editorial/ consistency	❏ No change
What changed		**Why the change**	
Revised this section to show that when a document is referenced in the standard (e.g. ISO 19011), it is referring to the most current version of the document unless otherwise specified.		This change was made to eliminate the need for years to be added to referenced standards. The changed language allows users to realize that the current version of a referenced standard is the one that the document is referring to unless otherwise specified.	
"ISO 9000:2000" was changed to "ISO 9000:2005."		This paragraph also updates the reference to ISO 9000. The year 2005 is specified in this case because at the time of publication of ISO 9001, it was undergoing revision.	

☑ Check for conformance to requirement
N/A

Title of clause/ subclause	3.0 Terms and definitions		
Intent of the text	This section provides definitions for terms that have a specific meaning within the standard that may or may not be identical to their generally accepted definition.		
Nature of the change	☑ Clarification/ correction	❑ Editorial/ consistency	❑ No change
What changed		**Why the change**	
Deleted the text of this clause relating to the transition of the supply chain from 1994 to 2000. This includes the explanation of supplier, organization, and customer.		When ISO 9001 was updated from the 1994 version to 2000, there was a significant change in basic terminology referring to customer, supplier, and organization. In the 2000 version, specific information was provided in 3.0 to clarify this. Because users have transitioned to the newer standard, this explanation is no longer required.	

☑ Check for conformance to requirement
N/A

Title of clause/ subclause	4.1 Quality management system General requirements		
Intent of the text	This section discusses the general requirements for the establishment, implementation, and maintenance of the QMS.		
Nature of the change	☑ Clarification/ correction	☐ Editorial/ consistency	☐ No change
What changed		**Why the change**	
Replaced "identify" with "determine" in clause (a).		Changed to "determine" due to consistency issues with the multiple uses of the terms "identify," "identification," and "identity" and the meaning of those words.	
Added "where applicable" after "measure" in clause (e).		Clarify that monitoring and measuring may not be applicable in all cases.	
In the last paragraph revised controls to type and extent of control to be applied.		The use of the phrase "type and extent of control" improves consistency with clause 7.4.	
Deleted the word "should" in the note.		The word "should" indicates that there is a requirement. Because notes are not to include requirements, the word "should" was deleted to clarify this.	
Added "analysis and improvement" in the note.		This change was made to maintain consistency with the name of clause 8.0.	
Added note 2 to explain that outsourced processes are those needed by the organization but		Leveraged information in guidance document on outsourced processes. In addition, increased	

What changed	Why the change
that the organization chooses to be performed by an external party.	emphasis that outsourced processes can be performed by any external party, not just those that are purchased.
Added note 3 to explain that the amount of control the organization is applying to the outsourced process is influenced by details such as the impact of the outsourced process on overall product realization in the organization, any shared control with the organization and the external party, controlling the process through clause 7.4. Note 3 also emphasizes that outsourcing does not absolve the organization of responsibility.	Clarify what is meant by the type and extent of control over outsourced processes.

☑ Check for conformance to requirement

❑ Determine processes within the QMS based on the scope of certification (reference 1.2).
❑ Develop the sequence and interaction of these processes.
❑ Decide on the criteria and methods for ensuring that processes are effective.
❑ Commit to providing the necessary resources for processes to function as specified.
❑ Monitor and measure processes.
❑ Continually improve processes.
❑ Ensure that outsourced processes are controlled.
❑ Verify that implementation is consistent with the process approach (reference 0.2).

Title of clause/ subclause	Documentation requirements 4.2.1 General		
Intent of the text	This subclause identifies the requirements for documentation.		
Nature of the change	☑ Clarification/ correction	☐ Editorial/ consistency	☐ No change
What changed		**Why the change**	
Added the word "records" to paragraphs (c) and (d). Deleted (e), which specifically addressed records as type of documentation.		This change is to emphasize the difference between documents and records.	
Revised to include words to show that the documents and records to be controlled are those that the organization determines to be necessary.		Organizations shall determine which documents and records are necessary to control.	
Revised note 1 to provide guidance that procedures can be combined into a single document or covered by more than one document.		This change was made to reflect the flexibility of the organization to develop procedures as unique documents or one document that combines more than one requirement.	

☑ Check for conformance to requirement

☐ To conform to this requirement, you must have:
  ☐ Documented statement of a quality policy
  ☐ Documented statement of quality objectives
  ☐ A quality manual
  ☐ Documented procedures (A minimum of six procedures are required.)
  ☐ Other documents and records needed to ensure the fulfillment of the requirements of your QMS

Title of clause/ subclause	Documentation requirements 4.2.2 Quality manual		
Intent of the text	This subclause describes the requirements for what must be included in an organization's quality manual.		
Nature of the change	☑ Clarification/ correction	❑ Editorial/ consistency	❑ No change
What changed		Why the change	
N/A		N/A	

☑ Check for conformance to requirement

❑ Your quality manual must:
  ❑ Define the scope of your QMS.
  ❑ Include justification for any exclusions in clause 7.
  ❑ Contain references to the documented procedures (must include at a minimum the required six procedures).
  ❑ Describe the interaction between the processes in the QMS.

Title of clause/ subclause	Documentation requirements 4.2.3 Control of documents		
Intent of the text	The subclause describes the requirements for the approval, review, revision, and maintenance of documents (regardless of media), including documents of external origin.		
Nature of the change	☑ Clarification/ correction	☐ Editorial/ consistency	☐ No change
What changed		Why the change	
Revised clause (f) to indicate the documents of external origin that are determined to be necessary by the organization are those that need to be controlled.		To clarify that only those external documents that are determined to be necessary by the organization for the planning and operation of the quality management system are included in this requirement.	

☑ Check for conformance to requirement

☐ You must have a document procedure for the control of documents. It must include provisions for:
  ☐ Documents approved prior to issue
  ☐ Review, update, and re-approval
  ☐ Changes known and current revision identified
  ☐ Documents available to those who need them
  ☐ Documents that are legible and identifiable
  ☐ Identifying and controlling documents of external origin
  ☐ Preventing the unintended use of obsolete documents and to identify them suitably if they are retained

Title of clause/ subclause	4.2.4 Control of records		
Intent of the text	This subclause describes the requirements to have a process and a documented procedure established to maintain records.		
Nature of the change	☑ Clarification/ correction	❑ Editorial/ consistency	❑ No change
What changed		Why the change	
Editorial changes that included restructuring the clause.		Changes in the structure and wording were made to bring closer compatibility with the same requirement for records found in ISO 14001. Wherever practical, common text and terminology was reviewed with ISO 14001. The focus was also changed to maintaining records.	

☑ Check for conformance to requirement

❑ Records established for the QMS must be controlled.
❑ You must have a documented procedure for the control of records. It must include provisions for:
  ❑ Identification of records required by your QMS or a reference to them
  ❑ Legibility
  ❑ Retrieval
  ❑ Storage
  ❑ Protection
  ❑ Retention time
  ❑ Disposition

Title of clause/ subclause	Management responsibility 5.1 Management commitment		
Intent of the text	This clause describes the need for top management to be directly involved in and to establish responsibility for the QMS.		
Nature of the change	☐ Clarification/ correction	☑ Editorial/ consistency	☐ No change
What changed		Why the change	
N/A		N/A	

☑ Check for conformance to requirement

☐ Top management must be directly involved in and establish responsibility for the QMS, including:
  ☐ Communicating to the organization the importance of meeting customer and statutory requirements
  ☐ Establishing the quality policy
  ☐ Establishing quality objectives
  ☐ Actively participating in management reviews
  ☐ Ensuring the availability of resources

Title of clause/ subclause	Management responsibility 5.2 Customer focus		
Intent of the text	This subclause describes the need for top management to ensure that customer requirements are met.		
Nature of the change	☐ Clarification/ correction	☐ Editorial/ consistency	☑ No change
What changed		Why the change	
N/A		N/A	

☑ Check for conformance to requirement
☐ Top management must ensure that customer requirements are determined and met, and that the organization is taking action to enhance customer satisfaction.

Title of clause/ subclause	Management responsibility 5.3 Quality policy		
Intent of the text	This clause describes the requirements for top management to establish the quality policy.		
Nature of the change	❑ Clarification/ correction	❑ Editorial/ consistency	☑ No change
What changed		**Why the change**	
N/A		N/A	

☑ Check for conformance to requirement
The quality policy must: ❑ Be appropriate to your organization.   ❑ Include a commitment to comply with requirements.   ❑ Include a commitment to continual improvement   ❑ Provide a framework for establishing and reviewing objectives.   ❑ Be communicated and understood throughout the organization.   ❑ Be reviewed for continued suitability.

Title of clause/ subclause	Quality planning 5.4.1 Quality objectives		
Intent of the text	This subclause describes the requirement for top management to ensure that quality objectives are established.		
Nature of the change	❏ Clarification/ correction	❏ Editorial/ consistency	☑ No change
What changed		Why the change	
N/A		N/A	

☑ Check for conformance to requirement
❏ Top management must establish quality objectives to meet product requirements. These objectives must be: ❏ Established at relevant functions and levels ❏ Measurable ❏ Consistent with the quality policy

Title of clause/ subclause	Quality planning 5.4.2 Quality management system planning		
Intent of the text	This subclause establishes top managements involvement in the planning for the QMS.		
Nature of the change	☐ Clarification/ correction	☐ Editorial/ consistency	☑ No change
What changed		Why the change	
N/A		N/A	

☑ Check for conformance to requirement

Top management ensures:
☐ Planning to carry out the requirements of QMS
☐ Planning to carry out the quality objectives
☐ The integrity of the QMS when things change

Title of clause/ subclause	Responsibility, authority and communication 5.5.1 Responsibility and authority		
Intent of the text	This subclause ensures that top management defines the responsibility and authority for the QMS.		
Nature of the change	☐ Clarification/ correction	☐ Editorial/ consistency	☑ No change
What changed		Why the change	
N/A		N/A	

☑ Check for conformance to requirement
☐ Top management must clearly define and communicate the responsibility and authority for the QMS

Title of clause/ subclause	Responsibility, authority, and communication 5.5.2 Management representative		
Intent of the text	This subclause describes the establishment and responsibility of the management representative.		
Nature of the change	☑ Clarification/ correction	❏ Editorial/ consistency	❏ No change
What changed		**Why the change**	
Added the words "the organization's" to management in the first paragraph.		To clarify that the management representative shall be a person of the organization's own management.	

☑ Check for conformance to requirement

❏ Top management must appoint a management representative. The management representative must:
❏ Ensure the QMS is established and implemented.
❏ Ensure the QMS is maintained.
❏ Report to top management on the performance of the QMS and needs for improvement.
❏ Promote the awareness of customer requirements throughout the organization.

Title of clause/ subclause	Responsibility, authority and communication 5.5.3 Internal communication		
Intent of the text	This subclause describes the requirement for top management to ensure that there is an appropriate communication process.		
Nature of the change	❑ Clarification/ correction	❑ Editorial/ consistency	☑ No change
What changed		Why the change	
N/A		N/A	

☑ Check for conformance to requirement
❑ Top management must ensure that an appropriate and effective communication process is established regarding the QMS.

Title of clause/ subclause	Management review 5.6.1 General
Intent of the text	This subclause describes that top management shall ensure that effective management reviews will take place.

Nature of the change	❑ Clarification/ correction	❑ Editorial/ consistency	☑ No change

What changed	Why the change
N/A	N/A

☑ Check for conformance to requirement

❑ Top management must ensure that effective management reviews are conducted to ensure that the QMS is suitable, adequate and effectiveness. The reviews should address:
  ❑ Opportunities to improve
  ❑ The need for change
  ❑ The quality policy
  ❑ The quality objectives
  ❑ In addition, records of management reviews must be kept.

suitable
adequate
effective

Title of clause/ subclause	Management review 5.6.2 Review input		
Intent of the text	This subclause describes the required inputs to management review.		
Nature of the change	❑ Clarification/ correction	❑ Editorial/ consistency	☑ No change
What changed		**Why the change**	
N/A		N/A	

☑   Check for conformance to requirement

The management review process includes examination of the following topics:
- ❑ Audits
- ❑ Customer feedback
- ❑ Process performance
- ❑ Product conformance
- ❑ Preventive actions
- ❑ Corrective actions
- ❑ Follow-up action from previous reviews
- ❑ Changes that could affect the QMS
- ❑ Recommendations for improvement

Title of clause/ subclause	Management review 5.6.3 Review output		
Intent of the text	This subclause describes the required outputs from the management review.		
Nature of the change	☐ Clarification/ correction	☐ Editorial/ consistency	☑ No change
What changed		**Why the change**	
N/A		N/A	

☑ Check for conformance to requirement

Records of management review must reflect decision and action items relative to:
☐ Improving the effectiveness of the QMS
☐ Improving processes
☐ Product improvement, especially as it relates to customer requirements
☐ Resource requirements

Title of clause/ subclause	Resource management 6.1 Provision of resources		
Intent of the text	This subclause describes the requirement for the organization to determine and provide needed resources.		
Nature of the change	☐ Clarification/ correction	☐ Editorial/ consistency	☑ No change
What changed		**Why the change**	
N/A		N/A	

☑ Check for conformance to requirement

The organization must determine the resources required to:
❑ Implement and maintain the QMS
❑ Continually improve the QMS
❑ Enhance customer satisfaction by meeting requirements

Title of clause/ subclause	Human resources 6.2.1 General		
Intent of the text	This subclause describes the requirement to have competent personnel.		
Nature of the change	☑ Clarification/ correction	☐ Editorial/ consistency	☐ No change

What changed	Why the change
Changed "product quality" to "conformity to product requirements."	This improves the consistency within ISO 9001 with the use of the phrases related to product quality and product requirements. This is the same phrase that is used in 4.1, paragraph 4.
Added a note that indicates conformity to product requirements may be affected by personnel who are indirectly or directly responsible for tasks within the quality management system.	Clarifies who the organization needs to apply this clause to.

☑ Check for conformance to requirement
☐ The organization must determine the competencies required for personnel who affect the QMS.   ☐ These competencies shall be based on education, skills, training, and experience.

Title of clause/ subclause	Human resources 6.2.2 Competence, training, and awareness		
Intent of the text	This subclause describes the requirements for ensuring the competence of employees. It also emphasizes that employees need to be aware of their responsibilities in the QMS.		
Nature of the change	☑ Clarification/ correction	❏ Editorial/ consistency	❏ No change

What changed	Why the change
The title wording order was changed from "Competence, awareness, and training" to "Competence, training, and awareness."	The change was made to reflect the structure of the wording in the clause that follows.
Revised to delete the words "product quality" in subclause (a).	Improves consistency within ISO 9001 with the use of the phrases relating to product quality and product requirements. This is the same phrase that is used in 4.1, paragraph 4.
Revised to add "where applicable" in subclause (b).	Provide the organization the option of determining that training or other actions might not be needed.
Revised the sentence to address necessary competence versus the satisfaction of needs in subclause (b).	Reduce ambiguity related to "satisfy these needs." Necessary competence is consistent with clause (a).

☑ Check for conformance to requirement
The organization must: ❏ Determine the criteria for competence of personnel. ❏ Where applicable, provide training or take other actions. In addition, the organization must:

☑️ Check for conformance to requirement
❑ Ensure that competence is achieved. ❑ Ensure that people are aware of the importance of their work and how they contribute to the fulfillment of objectives. ❑ Maintain records of education training, skills, and experience.

Title of clause/ subclause	Resource management 6.3 Infrastructure		
Intent of the text	This clause describes the requirements to provide an adequate infrastructure.		
Nature of the change	☐ Clarification/ correction	☑ Editorial/ consistency	☐ No change
What changed		Why the change	
Added information systems as an example in clause (c).		Broadens the examples without adding a requirement. Emphasizes the high dependency that organizations have on information systems.	

☑ Check for conformance to requirement

The organization must determine and provide applicable infrastructure needed for:
☐ Buildings, workspace, and utilities
☐ Equipment, including software
☐ Communication systems
☐ Transportation (e.g., a fleet of trucks)
☐ Information systems

Title of clause/ subclause	Resource management 6.4 Work environment		
Intent of the text	This clause describes the requirements for determining and managing the work environment.		
Nature of the change	☐ Clarification/ correction	☑ Editorial/ consistency	☐ No change
What changed		**Why the change**	
Added note to provide examples. Examples include physical, noise, temperature, humidity, and weather.		Explain what was meant by "work environment" to distinguish it from other uses of the word "environment" such as in ISO 14001. This avoids adding a term to ISO 9000.	

☑ Check for conformance to requirement
☐ The organization must determine and manage the work environment to achieve conformity to product requirements.

Title of clause/ subclause	Product realization 7.1 Planning of product realization		
Intent of the text	This clause requires planning for the needed processes for product realization.		
Nature of the change	☑ Clarification/ correction	❑ Editorial/ consistency	❑ No change
What changed		**Why the change**	
Editorial change to subclause (b).		Present wording is confusing. The second part suggests resources must be provided.	
Added "measurement" to subclause (c).		The word "measurement" was added for consistency of the use of the words at "monitoring and measurement" in other sections of ISO 9001.	

☑ Check for conformance to requirement

The organization's planning must include provisions for the appropriate:
❑ Quality objectives for the product
❑ Requirements consistent with processes determined in clause 4.1
❑ Processes
❑ Documents
❑ Resources
❑ Required verification, validation, monitoring, measurement, inspection, and testing
❑ Criteria for product acceptance
❑ Identification of records to provide evidence that product meets requirements

Title of clause/ subclause	Customer-related processes 7.2.1 Determination of requirements related to the product		
Intent of the text	This subclause describes the need for an organization to adequately determine product requirements.		
Nature of the change	☑ Clarification/ correction	❏ Editorial/ consistency	❏ No change
What changed		**Why the change**	
Revised "related" to "applicable" in clause (c).		Revised to remove ambiguity relating to nonproduct requirements that can be termed "related to the product."	
Revised "determined" to "considered necessary" in clause (d).		Clarifies that these requirements are different than the requirements identified in (a), (b), and (c).	
Added a note that describes what are considered post-delivery activities. These include warranty provisions, contractual obligations, and maintenance services.		The note clarifies what is meant by post-delivery activities without adding a definition to ISO 9000.	

☑ Check for conformance to requirement
The organization must have a process to determine the requirements related to its products, including the following requirements: ❏ Customer-specified ❏ Delivery ❏ Post-delivery ❏ Not stated but necessary for the intended use ❏ Statutory and regulatory ❏ Additional requirements considered necessary

Title of clause/ subclause	Customer-related processes 7.2.2 Review of requirements related to the product		
Intent of the text	This subclause describes the requirements for reviewing product requirements prior to contract acceptance.		
Nature of the change	☐ Clarification/ correction	☐ Editorial/ consistency	☑ No change
What changed		**Why the change**	
N/A		N/A	

☑   Check for conformance to requirement

The organization must have processes for reviewing product requirements prior to contract acceptance, including:
- ❑ Requirements are defined
- ❑ Resolving contract differences
- ❑ Confirming the organizations capability to produce the desired products
- ❑ Maintaining records of the review and the results associated with the review
- ❑ Confirming undocumented requirements prior to acceptance
- ❑ Establishing provisions for changes to requirements, including amending relevant documents and notifying personnel

Title of clause/ subclause	Customer-related processes 7.2.3 Customer communication		
Intent of the text	This subclause defines the requirements for communicating with customers.		
Nature of the change	☐ Clarification/ correction	☐ Editorial/ consistency	☑ No change
What changed		Why the change	
N/A		N/A	

☑ Check for conformance to requirement

The organization must have a process for communicating with customers about:
☐ Product information
☐ Inquiries
☐ Contracts
☐ Order handling
☐ Amendments to orders
☐ Customer feedback
☐ Complaints

Title of clause/ subclause	Design-and-development 7.3.1 Design-and-development planning		
Intent of the text	This subclause describes the requirements for the planning of design and development of product.		
Nature of the change	☑ Clarification/ correction	❑ Editorial/ consistency	❑ No change
What changed		**Why the change**	
Added a note that explains that review, verification, and validation and their subsequent records can be conducted separately or at the same time.		Clarifies the fact that review, verification, and validation can be conducted separately or at the same time.	

☑ Check for conformance to requirement
The organization must plan the design and development by doing the following: ❑ Determine design-and-development stages. ❑ Determine appropriate review, verification, and validation for each stage. ❑ Determine responsibilities and authorities for design and development. ❑ Manage the interface between groups to ensure effective communication and clear assignment of responsibility. ❑ Update plans as appropriate as the design-and-development stage progresses.

Title of clause/ subclause	Design-and-development 7.3.2 Design-and-development inputs		
Intent of the text	This subclause describes the required inputs to design and development.		
Nature of the change	☑ Clarification/ correction	☐ Editorial/ consistency	☐ No change
What changed		Why the change	
In the last paragraph changed "These inputs" to "The inputs."		Clarify that the list of inputs is what shall be included. The word "the" broadens that the list of inputs are not limited to items on the list.	

☑ Check for conformance to requirement

The organization must have a process in place to determine the required inputs to product design and development. These inputs should address:
❏ Function and performance
❏ Statutory and regulatory requirements
❏ Information from previous similar designs, where applicable
❏ Additional essential requirements
❏ Review inputs for adequacy.
❏ Ensure that these requirements are complete, unambiguous, and not in conflict with others.

Title of clause/ subclause	Design-and-development 7.3.3 Design-and-development outputs		
Intent of the text	This subclause describes the required outputs from the design-and-development process.		
Nature of the change	☑ Clarification/ correction	❑ Editorial/ consistency	❑ No change
What changed		**Why the change**	
Revised the first paragraph from "provided in a form that enables" to "shall be in a form suitable for."		Clarifies the sanctioned interpretation regarding the phrase "in a form." This choice of phrasing caused confusion in translating, in that the use of the word "form" was requiring that the outputs be in paper form.	
Added a note to indicate that preservation of product needs to be considered in the design-and-development outputs.		The note addresses the sanctioned interpretation regarding whether preservation of product needs to be considered in the design and development outputs.	

☑ Check for conformance to requirement
Outputs from the design process, as appropriate, shall: ❑ Meet the input requirements. ❑ Provide information for purchasing. ❑ Provide information for production and service provision ❑ Have product acceptance criteria. ❑ Specify characteristics essential for safe and proper use of product.

Title of clause/ subclause	Design and development 7.3.4 Design-and-development review		
Intent of the text	This subclause describes the requirements for appropriate reviews during the design process.		
Nature of the change	❑ Clarification/ correction	❑ Editorial/ consistency	☑ No change
What changed		**Why the change**	
N/A		N/A	

☑ Check for conformance to requirement
Design reviews must be: ❑ Conducted at suitable stages in accordance with plans ❑ Include representatives from functions concerned ❑ Evaluate the result of the design to meet requirements ❑ Used to identify problems and propose action ❑ In addition, records of review and necessary actions must be maintained

Title of clause/ subclause	Design and development 7.3.5 Design-and-development verification		
Intent of the text	This subclause describes the requirements for the verification of design-and-development activities.		
Nature of the change	❑ Clarification/ correction	❑ Editorial/ consistency	☑ No change
What changed		Why the change	
N/A		N/A	

☑ Check for conformance to requirement
❑ Verification must be performed to planned arrangements to: ❑ Ensure that outputs have met design-and-development input requirements. ❑ Records of verification results must be maintained.

Title of clause/ subclause	Design and development 7.3.6 Design-and-development validation		
Intent of the text	This subclause describes the requirements for the validation of design-and-development activities.		
Nature of the change	❑ Clarification/ correction	❑ Editorial/ consistency	☑ No change
What changed		Why the change	
N/A		N/A	

☑ Check for conformance to requirement

Design and development validation shall be conducted to planned arrangements to:
❑ Ensure that the resulting product is capable of meeting requirements for the specified application or intended use.
❑ Records of validation results and necessary actions must be maintained.
❑ Where practical, validation must be completed prior to delivery or implementation.

Title of clause/ subclause	Design and development 7.3.7 Control of design-and-development changes		
Intent of the text	This subclause describes the requirements for controlling changes identified during the design-and-development process.		
Nature of the change	❑ Clarification/ correction	❑ Editorial/ consistency	☑ No change
What changed		Why the change	
N/A		N/A	

☑   Check for conformance to requirement

The organization must ensure that design-and-development changes are:
- ❑ Reviewed
- ❑ Verified
- ❑ Validated
- ❑ Approved
- ❑ Evaluated to determine their effect on constituent parts
- ❑ In addition, records of changes and resulting actions must be maintained.

Title of clause/ subclause	Purchasing 7.4.1 Purchasing process		
Intent of the text	This subclause describes the requirements for the evaluation and selection of suppliers.		
Nature of the change	❑ Clarification/ correction	❑ Editorial/ consistency	☑ No change
What changed		Why the change	
N/A		N/A	

☑ Check for conformance to requirement

The organization must ensure purchased product conforms to requirements.
❑ The organization must consider the effect purchased product has on final product when evaluating suppliers, including the criteria for:
❑ Selection
❑ Evaluation
❑ Re-evaluation
❑ The organization must also determine the supplier's ability to supply product in accordance with requirements.
❑ Records of evaluations and actions taken must be maintained.

Title of clause/ subclause	Purchasing 7.4.2 Purchasing information		
Intent of the text	This subclause describes requirements for information to be contained in purchasing documents.		
Nature of the change	❏ Clarification/ correction	❏ Editorial/ consistency	☑ No change
What changed		**Why the change**	
N/A		N/A	

☑ Check for conformance to requirement

Information on purchase orders, as appropriate, must include:
- ❏ A description of the product or service
- ❏ A requirement for approval of products, procedures, processes, and equipment
- ❏ Requirements for qualification of personnel
- ❏ QMS requirements
- ❏ In addition, these requirements must be reviewed for adequacy prior to communicating with the supplier.

Title of clause/ subclause	Purchasing 7.4.3 Verification of purchased product		
Intent of the text	This subclause describes the requirements for the verification of conformance of purchased product.		
Nature of the change	❑ Clarification/ correction	❑ Editorial/ consistency	☑ No change
What changed		**Why the change**	
N/A		N/A	

☑ Check for conformance to requirement
❑ The organization must perform inspection or other activities to ensure conformance of purchased product. ❑ If appropriate, ensure provisions for making arrangements to perform product verification at supplier's location.

Title of clause/ subclause	Production and service provision 7.5.1 Control of production and service provisions		
Intent of the text	This subclause describes the controls placed on the production and service processes.		
Nature of the change	☑ Clarification/ correction	☐ Editorial/ consistency	☐ No change
What changed		Why the change	
Changed "devices" to "equipment" in clause (d).		"Devices" has been changed to "equipment" throughout the standard. Equipment is defined in ISO 9000 to include measuring instruments, which includes devices.	
Revised clause (f) by changing the word "release" to "product release."		Eliminates the need for the word device to be defined. The word "product" has been added for consistency within the clause.	

☑   Check for conformance to requirement
The organization must have controls in place for production and services processes regarding: ☐ The availability of product specifications ☐ Work instructions, as necessary ☐ Suitable equipment ☐ Use of monitoring-and-measuring equipment ☐ The implementation of monitoring and measurement ☐ Product release, delivery, and post-delivery activities

Title of clause/ subclause	Production and service provision 7.5.2 Validation of processes and production service provision		
Intent of the text	This subclause describes the requirements for validating processes whereby deficiencies cannot be verified.		
Nature of the change	☑ Clarification/ correction	☐ Editorial/ consistency	☐ No change
What changed		**Why the change**	
Revised first paragraph by making editorial changes to emphasize what processes this clause is referring to. The wording was changed to "as a consequence deficiencies are apparent only after the product is in use or the service has been delivered."		The word "this" was removed, because it was not clear what the word was referring to. The phrase "As a consequence" was added for clarification to emphasize the process this clause applies to.	

☑ Check for conformance to requirement

The organization must establish arrangements for validating processes where the output cannot be verified by monitoring or measurement.
☐ Processes must achieve planned results.
☐ As applicable, arrangements for these processes shall include:
   ☐ Defined criteria for approval and review
   ☐ Approval of equipment
   ☐ Qualification of personnel
   ☐ Specific methods and procedures
   ☐ Records
   ☐ Revalidation

Title of clause/ subclause	Production and service provision 7.5.3 Identification and traceability		
Intent of the text	This subclause describes the requirements for identification and traceability throughout the product realization process.		
Nature of the change	☑ Clarification/ correction	❑ Editorial/ consistency	❑ No change

What changed	Why the change
Paragraph two was revised to add the phrase "throughout product realization" at the end of the sentence.	"Throughout product realization" was added to provide consistency as it is used previously in this clause.
Paragraph 3 was revised to remove the words "and record" and added the words "and maintain records."	Revised use of the word record as a noun and not as a verb for consistency.

☑ Check for conformance to requirement

❑ The organization must have a process that describes the requirement for identification and traceability throughout the product realization process, including:
❑ Identifying product suitably
❑ Determining product status with respect to monitoring and measuring
❑ Maintaining records of unique identification, as appropriate

Title of clause/ subclause	Production and service provision 7.5.4 Customer property		
Intent of the text	This subclause describes the requirements for the control of customer property.		
Nature of the change	☑ Clarification/ correction	❑ Editorial/ consistency	❑ No change
What changed		**Why the change**	
Revised the last sentence of the clause, which addresses records, with an editorial change.		Provide consistency to reference records similar to other clauses.	
The note was revised to address personal data.		Personal data was added to address the increased awareness of the need to protect this type of data.	

☑ Check for conformance to requirement
The organization must have a process for the control of customer-owned property. The process should include: ❑ Identification and verification ❑ Protecting and safeguarding customer-owned property ❑ A provision for reporting damage, loss, or other problems to the customer ❑ Maintaining records ❑ Customer property provided for use or incorporation into the product.

Title of clause/ subclause	Production and service provision 7.5.5 Preservation of product		
Intent of the text	This subclause describes the requirements for the preservation of product.		
Nature of the change	☑ Clarification/ correction	❑ Editorial/ consistency	❑ No change
What changed		Why the change	
Changed the first sentence from "conformity of product" to "product."		The words "conformity of" were deleted for clarification and consistency with other clauses. This improves the consistency with ISO 9001 with the use of the phrases relating to product quality and product requirements.	
In the second sentence replaced "this" with "as applicable."		As applicable was added to clarify, specifically for service organizations, that this clause may not be applicable to all organizations.	

☑ Check for conformance to requirement
The organization must ensure that product and components are properly: ❑ Identified ❑ Packed ❑ Processed for delivery ❑ Stored ❑ Preserved ❑ Protected

Title of clause/ subclause	7.6 Control of monitoring and measuring equipment		
Intent of the text	This subclause describes the requirements for the control of equipment used in monitoring and measuring product and process.		
Nature of the change	☑ Clarification/ correction	❏ Editorial/ consistency	❏ No change

What changed	Why the change
Changed the word "devices" to "equipment" in the title and the first paragraph.	"Devices" has been changed to "equipment" throughout the standard. Equipment is defined in ISO 9000 to include measuring instruments, which includes devices. Eliminates the need for the word device to be defined.
"See 7.2.1" was deleted in the first paragraph.	"See 7.2.1" was deleted based on the revision made to 7.1.(c) to include measurement as a part of planning.
Changed "be calibrated or verified," to "be calibrated or verified, or both" in subclause (a).	Clarify the interpretation regarding that at times calibration, verification, or both are required.
Changed "be identified" to "have identification" in subclause (c).	Consistent use of the words "identify" and "identification." The change of "be identified" to "have identification" clarifies that the status of equipment can be made in various methods.
Separated the requirement for records as a stand-alone paragraph.	Clarify that the records requirement did not just pertain to the paragraph it was attached to.

What changed	Why the change
Note referring to ISO 10012 was deleted.	Note referring to 10012 was confusing to users because it did not provide additional guidance for ISO 9001.
Added note that explains that confirmation of software includes verification and configuration to maintain its suitability for use.	Clarified the intent of verification of software.

☑ Check for conformance to requirement

To comply with this subclause's requirements, the organization must:
❏ Select appropriate equipment for monitoring and measuring.
❏ Calibrate and/or verify instruments at defined intervals.
❏ Calibrate using known/traceable standards.
❏ Identify calibration status.
❏ Safeguard unintended adjustments.
❏ Adjust and re-adjust equipment as necessary.
❏ Protect equipment from damage or invalidation.
❏ Establish provisions for assessing prior adjustments in the event a problem arises.
❏ Have a process for confirming software.
❏ Maintain calibration records.

Title of clause/ subclause	Measurement, analysis, and improvement 8.1 General		
Intent of the text	This clause describes the general requirements that are needed for the measurement, analysis, and improvement processes.		
Nature of the change	☐ Clarification/ correction	☑ Editorial/ consistency	☐ No change
What changed		**Why the change**	
Revised "product" to "product requirements" in subclause (a)		Improves consistency within ISO 9001 with the use of the phrases relating to product quality and product requirements. Use the same text as clause 4.1, paragraph four.	

☑ Check for conformance to requirement

The organization must have controlled processes for monitoring, measurement, analysis, and improvement processes for:
☐ Conformity of product
☐ Conformity of the QMS
☐ Continually improving the effectiveness of the QMS
☐ Determination and extent of use of applicable statistical methods

Title of clause/ subclause	Monitoring and measurement 8.2.1 Customer satisfaction		
Intent of the text	This subclause describes the requirements for determining methods for monitoring customer satisfaction.		
Nature of the change	☑ Clarification/ correction	☐ Editorial/ consistency	☐ No change
What changed		Why the change	
Added a note which identifies that sources of information for customer perception can be surveys, data on products delivered to customers, and compliments.		Clarifies what is meant by "methods for obtaining and using this information shall be determined."	

☑   Check for conformance to requirement

☐ The organization must monitor customer perception as to whether the organization has met customer requirements.
☐ In addition, the organization must determine methods for obtaining and using information.

Title of clause/ subclause	Monitoring and measurement 8.2.2 Internal audit		
Intent of the text	This subclause describes the requirements for conducting internal audits.		
Nature of the change	☑ Clarification/ correction	❑ Editorial/ consistency	❑ No change

What changed	Why the change
This clause was restructured.	Changed structure to improve flow and to clarify all of the listed actions that are required to be in the procedure. Clarifies the records to be maintained with a stand-alone sentence.
The last paragraph was revised to include necessary corrections in addition to corrective action.	Clarifies that correction can be made to nonconformities themselves in addition to corrective action.
The reference to ISO 10011 was revised to ISO 19011.	ISO 19011 was released after ISO 9001:2000 was published.

☑   Check for conformance to requirement

The organization must have a documented procedure for internal audits. This procedure must:
- ❑ Define responsibilities for planning and conducting audits.
- ❑ Establish records.
- ❑ Report results.
- ❑ Conduct internal audits at planned intervals.
- ❑ Comply with planned arrangements.
- ❑ Conform to the requirements of ISO 9001.
- ❑ Conform to QMS requirements established by the organization
- ❑ Be effectively implemented and maintained.

☑ Check for conformance to requirement

❏ The audit program shall include:
❏ Planned internal audits, taking into consideration the importance of area and results of previous audits.
❏ Defined audit criteria, scope, frequency, and method. Selection of auditors must be impartial. (Auditors may not audit their own work.)
❏ A commitment from management to undertake correction or corrective actions to correct nonconformances uncovered during audits.
❏ Follow-up activities to verify actions taken.
❏ A record of the results of past audits.

Title of clause/ subclause	Monitoring and measurement 8.2.3 Monitoring and measurement of processes		
Intent of the text	This process defines the requirements for establishing methods for determining what you are going to measure to determine if processes meet requirements.		
Nature of the change	☑ Clarification/ correction	☐ Editorial/ consistency	☐ No change
What changed		**Why the change**	
The clause was revised to delete the phrase "to ensure conformity of the product."		Differentiate between this clause and 8.2.4, Monitoring and measurement of product. Clarify the confusion around the phrase "to ensure conformity of the product."	
Added a note to emphasize that the type and extent of monitoring and measurements is dependent on the product's effect on the QMS.		Maintain the association to product without focusing on it in the text of the clause.	

☑ Check for conformance to requirement

The organization must:
- ☐ Establish methods for monitoring processes.
- ☐ Demonstrate, where applicable, measurement of the quality management processes.
- ☐ Use methods that demonstrate the ability of process to meet requirements.
- ☐ Use correction and corrective action if planned results aren't achieved.

Title of clause/ subclause	Monitoring and measurement 8.2.4 Monitoring and measurement of product		
Intent of the text	This clause defines the requirements for monitoring and measuring product in order to verify that requirements have been met.		
Nature of the change	☑ Clarification/ correction	❑ Editorial/ consistency	❑ No change
What changed		**Why the change**	
Restructured the paragraph to more clearly show that the records to be maintained are those that authorize the release of the product or delivery to the customer.		Structure of clause was changed to emphasize what records are required. Consistency of text within the clause.	

☑ Check for conformance to requirement

❑ Monitor and measure product characteristics to ensure that requirements have been met.
❑ Establish and implement methods at appropriate stages and according to planned arrangements.
❑ Retain evidence of conformity.
❑ Keep records that indicate persons authorizing release.
❑ Do not proceed with product release or service delivery unless planned arrangements have been met or otherwise approved by a relevant authority and, where applicable, by the customer.

Title of clause/ subclause	Monitoring and measurement 8.3 Control of nonconforming product		
Intent of the text	This clause describes the process for controlling nonconforming product.		
Nature of the change	☑ Clarification/ correction	❑ Editorial/ consistency	❑ No change

What changed	Why the change
Restructured clause.	Restructured clause for consistency within the standard and to make it more friendly for all organizations to apply.
Added the words "where applicable" to paragraph 2.	"Where applicable" was added to clarify that not all methods of controlling nonconforming product can be applied to every organization.
Moved the last paragraph of the clause that addressed the requirement of taking appropriate action to the list making it item (d).	Clarify for organizations where nonconforming product cannot be detected until after delivery or use has started that taking appropriate action is one of the methods that can be used to control nonconforming product.

☑ Check for conformance to requirement
❑ The organization must identify and control nonconforming product to prevent its unintended use or delivery.   ❑ The organization must have a documented procedure for controlling nonconforming product by defining the controls and related responsibilities and authorities.   ❑ Where applicable, nonconforming product is managed in one or more of the following ways:

☑ Check for conformance to requirement

❑ Take action to eliminate detected nonconformity.
❑ Authorize its use with concession by relevant authority or customer.
❑ Take action to prevent its unintended use.
❑ Take appropriate action if nonconforming product is detected after delivery or use has started.
❑ Nonconforming product is subject to re-verification after it is corrected.
❑ Records of nonconformities and actions taken, including consessions, are maintained.

Title of clause/ subclause	Monitoring and measurement 8.4 Analysis of data		
Intent of the text	This clause describes the requirements for analyzing data that demonstrate the suitability and effectiveness of the quality management system.		
Nature of the change	☑ Clarification/ correction	☐ Editorial/ consistency	☐ No change
What changed		**Why the change**	
Changed the reference in clause (b) from 7.2.1 to 8.2.4.		Correct the reference. Clause 7.2.1 is where requirements are determined where clause 8.2.4 is where conformity to product requirements is determined.	
Changed clause (c) to include references to 8.2.3 and 8.2.4.		Consistent use of references with other portions of this clause.	
Changed clause (d) to include a reference to 7.4.		Consistent use of references with other portions of this clause.	

☑ Check for conformance to requirement

☐ The organization must determine, collect, and analyze data generated from monitoring and measurement and other relevant sources relative to:
☐ Customer satisfaction
☐ Product conformity
☐ Characteristics and trends of products and processes
☐ Opportunities for preventive action
☐ Suppliers
☐ Data are evaluated to determine opportunities for continual improvement of the effectiveness of the QMS.

Title of clause/ subclause	Improvement 8.5.1 Continual improvement		
Intent of the text	This subclause describes the requirements for using information derived from the QMS to continually improve the organization.		
Nature of the change	❑ Clarification/ correction	❑ Editorial/ consistency	☑ No change
What changed		Why the change	
N/A		N/A	

☑ Check for conformance to requirement
The organization must continually improve through the use of the: ❑ Quality policy ❑ Quality objectives ❑ Audit results ❑ Analysis of data ❑ Corrective action ❑ Preventive action ❑ Management review

Title of clause/ subclause	Continual improvement 8.5.2 Corrective action		
Intent of the text	This subclause describes the requirements for taking corrective action to eliminate nonconformities to prevent recurrence.		
Nature of the change	☑ Clarification/ correction	❏ Editorial/ consistency	❏ No change
What changed		**Why the change**	
Revised the first paragraph from cause to "causes."		Clarify that nonconformities can have more than one cause.	
Revised to include "effectiveness" in clause (f).		Clarify that when corrective actions are reviewed the effectiveness of the actions is to be considered.	

☑   Check for conformance to requirement

❏ The organization must take action to eliminate cause of nonconformity to prevent recurrence.
❏ Corrective actions taken must be appropriate to effect of nonconformance encountered.

The documented procedure for corrective action must define requirements for:
❏ Reviewing nonconformances, including customer complaints.
❏ Determining cause.
❏ Evaluating need to take action to prevent recurrence.
❏ Determining and implementing action.
❏ Maintaining records of the results of actions taken.
❏ Reviewing effectiveness of corrective action taken.

Title of clause/ subclause	Improvement 8.5.3 Preventive action		
Intent of the text	This subclause describes the requirements for taking action to prevent the occurrence of potential nonconformities.		
Nature of the change	☐ Clarification/ correction	☐ Editorial/ consistency	☑ No change
What changed		Why the change	
Revised to include "effectiveness" in clause (f).		Clarify that when corrective actions are reviewed the effectiveness of the actions is to be considered.	

☑ Check for conformance to requirement

☐ The organization must take action to eliminate the causes of potential nonconformities.
☐ Ensure that preventive actions are appropriate to the effects of the potential problem.

The documented procedure for preventive action must:
☐ Determine potential problems and their cause.
☐ Evaluate need to take action to prevent occurrence.
☐ Determine and implement action.
☐ Maintain records of the results of actions taken.
☐ Review effectiveness of preventive actions taken.

ISO STANDARD SPEAK—
DECIPHERED

This appendix is intended to give the reader, user, novice, or all three, a fighting chance of working through the myriad of acronyms, numbers, and language that can bog down any technical writing. This isn't intended to be a glossary or list of standard definitions. It's merely an attempt to provide some common language and guidance for more information. The first recommendation for more information will start by directing you to the International Organization for Standardization's (ISO) Web site at *www.iso.org*.

INTERNATIONAL WORKSHOP AGREEMENT (IWA)

An international workshop agreement (IWA) is one of the potential deliverables or documents that are covered within the standards-development processes. However, an IWA is an ISO document that is developed through a meeting, or series of meetings, rather than through the formal technical committee process that has been described in this book. Therefore, these documents can be proposed and developed by interested stakeholders directly and not necessarily through a member body. Although they bare the ISO "brand," they are not requirement documents. For example, IWA 1:2005 Quality management systems—Guidelines for process improvements in health service organizations, was developed by, as the name implies, stakeholders from the health services industry. This was developed to give additional guidance to their user community in the application of QMS within the health services industry.

ISO 9000 INTRODUCTION AND SUPPORT PACKAGE

The following documents were developed to support and give guidance to the application of the various concepts given in the requirements within ISO 9001:2000. One of the goals of the amendment was to address some of the concepts illustrated in the guidance documents that needed clarity.

■ *Guidance on "Outsourced Processes" (N630R2).* This document provides guidance on the application of clause 4.1, in ISO 9001:2000, with regard to outsourced processes.

■ *Guidance on ISO 9001:2000 clause 1.2, Application (N524R).* This document provides guidance on the intent of subclause 1.2, Application, in ISO 9001:2000, and to give some examples of its use in typical situations.

■ *Guidance on the documentation requirements of ISO 9001:2000 (N525R).* This document provides guidance to the amount and detail of documentation that was required by ISO 9001:2000, given that one of the focuses of the 2000 revision was to reduce the requirements for documentation to a level that is consistent with the desired results that the organization and its customers expect.

■ *Guide to the terminology used in ISO 9000:2000 family of standards (N526R).* This document provides guidance on some of the terms that are used in the ISO 9000 family of documents, and refers the reader to the specific ISO documents which address the terms and definitions used.

■ *Guidance on the concept and use of process approach for management systems (N544R2).* This document provides guidance on the intent and the application of the concept of the "process approach" in ISO 9001:2000.

ISO 9001:2008

ISO 9001 is the fundamental document that this book refers to and the ":2008" refers to the year in which that revision was published.

ISO GUIDE 72

This guide is published by ISO and provides guidance on justifying the development of, or the continued relevance of, a management system standard, including a view into the market relevance of the document. It also provides guidance on the methods used to develop and maintain the documents themselves and the terminology that is used within them. Following the guidance of this document is the first step in the process of developing a standard as illustrated in chapter 5 of this book.

ISO/TC 176 AND ISO/TC 207

These technical committees (TCs) are groups of subject matter experts in a given field who act as a project team. They are responsible for the development and maintenance of content related to a given subject area on behalf of ISO. For example, ISO/TC 176 is responsible for the content related to quality management systems. (Visit ISO/TC 176's Web site at *www.tc176.org*.) ISO/TC 207 is responsible for environmental management systems. (Visit ISO/TC 207's Web site at *www.tc207.org*.)

ISO/TC 176 INTERPRETATIONS

A process for providing the user community a forum for officially sanctioned interpretations was created after the release of ISO 9001:2000. There is an official process by which a WG considers requests for interpretations under rather strict criteria for submission. The interpretations are then balloted by the member bodies to ISO/TC 176, and those that are accepted are officially documented and published. For additional information, visit the ISO/TC 176 Web site at *www. tc176.org/Interpre.asp*. The interpretations process was one of the significant sources that was used in the drafting of the amendment.

SC 1/SC 2/SC 3

These are the subcommittee structures that exist one tier down from the technical committees. As you might expect, these subcommittees are responsible for a subset of the work of the overall technical committee. ISO/TC 176's subcommittees have the following responsibilities:

- SC1—Concepts and terminology used in the documents related to QMS, including responsibility for ISO 9000
- SC 2—QMS, including responsibility for ISO 9001 and ISO 9004
- SC 3—Supporting technologies and documents related to QMS

TECHNICAL MANAGEMENT BOARD (TMB)

The ISO Technical Management Board (TMB) has been established to serve a governance role on the technical activities carried out within ISO. It has specific responsibilities to establish the technical committees and their leadership, as well as providing governance to the work of the technical committees through formally documented directives.

TECHNICAL SPECIFICATION (TS)

A technical specification is one of the potential deliverables of a work item within a technical committee. It's a requirements document that is achieved via consensus balloting by member nations as is done with other ISO documents. A technical specification must be reviewed every three years for continued applicability and must be confirmed for another three-year period, withdrawn, revised, or it must be elevated to the status of an international standard. After a six-year period, the technical specification is to be revised and published as an international standard or withdrawn. It should be noted, however, that ISO/TS 16949, the automotive sector-specific standard referred to in this book, has surpassed this mark. This was not an oversight. It was debated and was given additional latitude due to its market acceptance and utilization, and the ISO TMB extended its status beyond this requirement. It's not clear at the time of publication what the future will hold for ISO/TS 16949, but there appears to be no cause for concern on its continued viability, relevance, or use.

U.S. TECHNICAL ADVISORY GROUP TO ISO/TC 176

This represents the U.S. technical advisory group (TAG) to ISO/TC 176, and has been established to develop and express the U.S. position on the activities of this technical committee, including official ISO ballots on the various standards and documents developed within ISO/TC 176. This group is accredited through the American National Standards Institute (ANSI). More information on ANSI can be found at *www.ansi.org*. The American Society for Quality (ASQ) provides invaluable support to the U.S. TAG in the area of administration of its activities and membership. More information on ASQ can be found at *www.asq.org*.

WD/CD/DIS/FDIS

These acronyms depict the standards-development process that is covered in chapter 4 of this book, so we won't go into great detail here. They represent the progressive stages that a document goes through from concept to fully balloted and published. The acronyms themselves stand for:

- WD—Working draft
- CD—Committee draft
- DIS—Draft international standard
- FDIS—Final draft international standard

WG/TG

These acronyms refer to some of the levels of organizational structure found within the ISO TCs. The base level is the task group (TG). Within the TGs, specific work items, including the development of standards and guidance documents, are addressed by subject matter experts made up from the various member nations within the TC. For example, TG 1.19 is the task group that was charged with the development of the ISO 9001:2008 amendment. The working group (WG) level has an administrative and governance role for a collection of TGs working in the same subject area. For example, WG 18 is responsible for the TGs related to the drafting of the ISO 9001 and ISO 9004 standards.

INDEX

O

P

Q

R

S

T

V

W